PROBLEM SOLVING
A statistician's guide

Second edition

Chris Chatfield
Reader in Statistics, University of Bath, UK

CHAPMAN & HALL/CRC

Boca Raton   London   New York   Washington, D.C.

Library of Congress Cataloging-in-Publication Data

Catalog record is available from the Library of Congress

Visit the CRC Press Web site at www.crcpresss.com

© 1988, 1995 by Chris Chatfield
First edition 1988
Reprinted 1990, 1991, 1993
Second edition 1995
Originally published by Chapman & Hall

No claim to original U.S. Government works
International Standard Book Number 0-412-60630-5
Library of Congress Card Number 94-74702
Printed in the United States of America   3 4 5 6 7 8 9 0
Printed on acid-free paper

PROBLEM SOLVING
A statistician's guide

CHAPMAN & HALL TEXTS IN STATISTICAL SCIENCE SERIES

Editors:

Dr Chris Chatfield
Reader in Statistics
School of Mathematical Sciences
University of Bath, UK

Professor Jim V. Zidek
Department of Statistics
University of British Columbia
Canada

OTHER TITLES IN THE SERIES INCLUDE

Practical Statistics for Medical Research
D.G. Altman

Interpreting Data
A.J.B. Anderson

Statistical Methods for SPC and TQM
D. Bissell

Statistics in Research and Development
Second edition
R. Caulcutt

The Analysis of Time Series
Fourth edition
C. Chatfield

Statistics in Engineering – A Practical Approach
A.V. Metcalfe

Statistics for Technology
Third edition
C. Chatfield

Introduction to Multivariate Analysis
C. Chatfield and A.J. Collins

Modelling Binary Data
D. Collett

Modelling Survival Data in Medical Research
D. Collett

Applied Statistics
D.R. Cox and E.J. Snell

Statistical Analysis of Reliability Data
M.J. Crowder, A.C. Kimber,
T.J. Sweeting and R.L. Smith

An Introduction to Generalized Linear Models
A.J. Dobson

Introduction to Optimization Methods and their Applications in Statistics
B.S. Everitt

Multivariate Studies – A Practical Approach
B. Flury and H. Riedwyl

Readings in Decision Analysis
S. French

Multivariate Analysis of Variance and Repeated Measures
D.J. Hand and C.C. Taylor

The Theory of Linear Models
B. Jorgensen

Statistical Theory
Fourth edition
B. Lindgren

Randomization and Monte Carlo Methods in Biology
B.F.J. Manly

Statistical Methods in Agriculture and Experimental Biology
Second edition
R. Mead, R.N. Curnow and
A.M. Hasted

Elements of Simulation
B.J.T. Morgan

Probability: Methods and Measurement
A. O'Hagan

Essential Statistics
Third edition
D.G. Rees

Large Sample Methods in Statistics
P.K. Sen and J.M. Singer

Decision Analysis: A Bayesian Approach
J.Q. Smith

Applied Nonparametric Statistical Methods
Second edition
P. Sprent

Elementary Applications of Probability Theory
Second edition
H.C. Tuckwell

Statistical Process Control: Theory and Practice
Third edition
G.B. Wetherill and D.W. Brown

Statistics for Accountants
S. Letchford

Full information on the complete range of Chapman & Hall statistics books is available from the publishers.

Contents

PART THREE APPENDICES

Preface

There are many books on statistical theory and on specific statistical techniques, but very few which concern themselves with **strategies** for problem solving. This book is written for anyone who has studied a range of basic statistical topics but still feels unsure about tackling real-life problems. How can reliable data be collected to answer a specific question? What is to be done when confronted with a set of real data, perhaps rather 'messy' and perhaps with unclear objectives?

Problem solving is a complex process which is something of an acquired knack. This makes it tricky to teach. The situation is not helped by those textbooks which adopt a 'cookbook' approach and give the false impression that statistical techniques can be performed 'parrot-fashion'.

Part One of this book aims to clarify the general principles involved in tackling real-life statistical problems, while Part Two presents a series of exercises to illustrate the practical problems of real data analysis. These exercises are concerned with solving problems rather than with just using techniques – an important distinction. The book aims to develop a range of skills including getting a 'feel' for data, the ability to communicate and to ask appropriate searching questions and realizing the importance of understanding context and of avoiding trouble. It also demonstrates the exciting potential for simple ideas and techniques, particularly in the emphasis on the initial examination of data (or IDA).

In this revised second edition, I have substantially rewritten and extended the discussion in Part One of the book and have added further examples. In Part Two, I have added new exercises E.5, E.6 and G.5(b). The references have been updated throughout, including Appendix A.

This is essentially a practical book emphasizing general ideas rather than the details of techniques, although Appendix A provides a brief, handy reference source. Nevertheless, I want to emphasize that the statistician needs to know sufficient background theory to ensure that procedures are based on a firm foundation. Fortunately many teachers already present a good balance of theory and practice. However, there is no room for complacency as theory can unfortunately be taught in a counter-productive way. For example, a student who first meets the t-test as a special case of the likelihood ratio test may be put off statistics for life! Theory taught in the right way should strengthen practical judgement, but, even so, the essence of statistics is the collection and analysis of data, and so theory must be backed up with practical experience.

This book is based on a course I have given for several years at Bath University to final-year statistics undergraduates. As well as general lectures, a selection of exercises is written up in a practical book or as special projects, and the course is examined by continuous assessment. Student reaction suggests the course is well worthwhile in improving motivation and providing valuable practical experience.

In describing my approach to problem solving, I have tried to be truthful in saying what I would actually do based on my experience as a practising statistician. I hope I do not live to regret putting my 'confessions' into print! I recognize that the book 'will not please all the people all the time'. It is often said that ten statisticians will offer ten different ways of tackling any given problem. However if the book stimulates debate on general principles, provides readers with useful practical guidance, and encourages other statisticians to write similar books based on their alternative, and perhaps very different, experience, then I will regard the book as successful.

I am indebted to many people for helpful comments during the writing of this book. They include David Cox, David Draper, Andrew Ehrenberg, David Hand, David Hinkley, Elizabeth Johnston, Roger Mead, Bernard Silverman and David Williams, as well as all the many people who have given me advice, encouragement, real data and real problems over the years. Particular thanks go to Jim Hodges for Example 4.3(3). Of course any errors, omissions or obscurities which remain are probably (?) my responsibility. The author will be glad to hear from any reader who wishes to make constructive comments.

Finally, it is a pleasure to thank Mrs Sue Collins for typing the manuscript for the first edition with exceptional efficiency. I also thank my word processor for enabling me to type the revisions for this second edition myself!

Chris Chatfield
School of Mathematical Sciences
University of Bath
Bath BA2 7AY
e-mail: cc@maths.bath.ac.uk
July 1994

Prelude

What are statistical problems really like?

One real-life example will suffice at this stage to illustrate that many textbook examples are over-simplified and rather artificial. Your telephone rings. A doctor at a local hospital has collected some observations in order to compare the effects of four anaesthetic drugs, and wants some help in analysing the data. A meeting is arranged and the statistician is assured that 'it won't take long' (of course it does! but statisticians must be willing to respond to genuine requests for help). When the data are revealed, they turn out to consist of as many as 31 variables measured for each of 80 patients undergoing surgery for a variety of conditions, such as appendicitis. A small portion of the data is shown in Table P.1. The doctor asks 'How should the data be analysed?'. How indeed! In order to try and answer this question, I put forward six general rules (or rather guidelines).

Table P.1 Part of the anaesthetic data

Patient no.	3	5	14	27	42	etc.
Group	A	B	D	A	C	
Sex	M	F	F	F	F	
Age	38	38	42	41	54	
Operation	RIH	Mastectomy	TAH	Laparotomy	Appendix	
Premed time	?	?	116	90	59	
Vapour	/	/	/	HAL	/	
T_1	3	4	3	10	1	
T_2	3	4	3	10	1	
T_3	48	?	110	108	38	
Antiemetic	?	200	25	60	/	
Condition	2	2	1	3	1	
etc.						

Rule 1 Do not attempt to analyse the data until you understand what is being measured and why. Find out whether there is any prior information about likely effects.

You may have to ask lots of questions in order to clarify the objectives, the meaning of each variable, the units of measurement, the meaning of special symbols, whether similar experiments have been carried out before, and so on.

Here I found the following:

1. The group (A, B, C or D) is used to denote the anaesthetic administered, and the main objective is to compare the effects of the four drugs.

2. A question mark denotes a missing observation.
3. A slash (/) denotes NO or NONE (e.g. no vapour was inhaled by patient 3).
4. T_1 denotes time (in minutes) from reversal of anaesthetic till the eyes open, T_2 the time to first breath, and T_3 the time till discharge from the operating theatre.
5. Overall condition after the operation is rated from 1 (very good) to 4 (awful).
6. And so on...

The doctor was unaware of any previous experiments to compare the four anaesthetics and wanted the data analysed as a 'one-off' sample. It seems unlikely that the prior information was so meagre and this point may be worth pursuing if the results are inconclusive.

Rule 2   Find out how the data were collected.

In this case, were patients allocated randomly to the four groups? How important is the type of operation as far as after-effects were concerned? How reliable are the measurements? etc. If the experiment has not been properly randomized, then a simple descriptive analysis may be all that can be justified.

Rule 3   Look at the structure of the data.

Are there enough observations? Are there too many variables? What sort of variables are they?

Here there were 20 patients in each of the four groups, and this should be enough to make some sort of comparison. The number of variables is high and so they should not all be considered at once. Try to eliminate some in consultation with the doctor. Are they really all necessary? For example, T_1 and T_2 were usually identical and are essentially measuring the same thing. One of them should be excluded.

It is helpful to distinguish the different types of variable being studied. Here the variables generally fall into three broad categories. There are demographic variables (e.g. age) which describe each patient. There are controlled variables (e.g. type of vapour) which are generally under the control of hospital staff. Finally, there are response variables (e.g. condition) which measure how the patient responds to the operation. Variables can also be usefully classified by the type of measurement as continuous (e.g. time), discrete (e.g. number of children), qualitative (e.g. the type of operation), binary (e.g. male or female), etc. The ensuing analysis depends critically on the data structure.

Rule 4   The data then need to be carefully examined in an exploratory way, before attempting a more sophisticated analysis.

The given data are typical of many real-life data sets in that they are some-what 'messy', with missing observations, outliers, non-normal distributions as well as a mixture of qualitative and quantitative variables. There are various queries about the quality of the data. For example, why are some observations missing, and are there any obvious errors? The initial examination of the data should then continue by calculating appropriate summary statistics and plotting the data in whatever way seems appropriate. Do this for each of the four groups separately.

First look at the demographic variables. Are the groups reasonably comparable? Try to avoid significance tests here, because there are several variables and it is the overall comparison which is of interest. Second, look at the controlled variables. Were the same surgeon and anaesthetist involved? Is there evidence of different strategies in different groups? Third, look at the response variables. Examine them one at a time, at least to start with (e.g. Exercise B.9). Here a one-way ANOVA may be helpful to see if there are important-looking differences between the group means, but don't get too concerned about 'significance' and the size of P-values at this stage. With several variables to examine, different operations involved, and doubts about randomization, it is more important to see if any differences between groups are of practical importance and to see if this fits in with any prior knowledge. Only after this initial analysis will it be possible to see if any further, perhaps more complicated, analysis is indicated.

Rule 5   Use your common sense at all times.

Rule 6   Report the results in a clear, self-explanatory way.

These general principles will be amplified and extended in Part One of this book, while Part Two presents a series of worked exercises to illustrate their use.

How to tackle statistical problems

A BRIEF SUMMARY

Understanding the problem

What are the objectives? What background information is available? Can you formulate the problem in statistical terms? Ask Questions.

Collecting the data

Have the data already been collected? If so, how? If not, should an experimental design, sample survey, observational study, or what, be used? How will randomization be involved?

Analysing the data

Process the data. Check data quality. Carry out an initial examination of the data (an IDA). Are the conclusions then obvious? If not, has the formal method of analysis been specified beforehand? If so, does it still look sensible after seeing the data? If not, how do we select an appropriate method of analysis?

Have you tackled a similar problem before? If not, do you know someone else who has? Or can you find a similar problem in a book? Can you restate the problem or solve part of the problem? Is there prior information (empirical or theoretical) about a sensible model, and, if so, do the new results agree with it? Would the analysis be easier if some variables were transformed, or if some non-standard feature were removed (and is this justifiable)? Try more than one analysis if unsure (e.g. with or without an outlier; parametric or non-parametric approach) and see if the results are qualitatively different. Ask for help if necessary. Check any model that you fit, not only with a residual analysis, but also by replicating the results where possible.

Presenting the results

Are the conclusions what you expected? If they are counter-intuitive, have you performed an appropriate analysis? Do you have to write up a report? Plan the structure of your report carefully. Revise it several times. Ask a friend to read it before you hand it over.

PART ONE

The General Principles Involved in Tackling Real-life Statistical Problems

This part of the book gives general advice on tackling real-life statistical problems. In addition to material on problem formulation and the collection of data, the text also covers some important topics, such as using a library and report writing, which are indispensable to the applied statistician but which may not be covered in conventional statistics courses.

The reader is assumed to have a working knowledge of simple probability models and basic inference. Detailed techniques are not discussed here, although a brief digest is given in Appendix A. As the computer takes over the routine implementation of many techniques, it is arguable that it will become less important to remember all their details. Instead the analyst will be able to concentrate on general strategy such as selecting the most appropriate method of analysis, checking assumptions and interpreting the results. It is this general strategy which is the main substance of this book.

1
Introduction

1.1 What is statistics?

Statistics is concerned with **collecting**, **analysing** and **interpreting** data in the best possible way, where the meaning of 'best' depends on the particular circumstances of the practical situation.

Note the three different facets of a typical statistical study. Statistics is much more than just analysing data.

A key element of statistical problems is the presence of **uncertainty** and **inherent variability** in the measurements. Despite this uncertainty, the statistician has to make recommendations, though whether statistics should include **decision-making** is a matter of controversy. Barnett (1982, Chapter 1) discusses possible definitions of statistics, but there is no simple answer with which everyone will agree.

Another key element of statistics is the use of **inductive** reasoning. This involves arguing from the particular (a body of data) to the general (a model or theory for future behaviour). This contrasts with the **deductive** nature of probability theory (and of mathematics in general), wherein one argues from a given model or from known axioms to make statements about the behaviour of individuals or about samples drawn from a known population. In practice (section 7.4.1) scientific progress usually involves a combination of inductive and deductive reasoning.

1.2 The statistician as problem solver

A statistician needs to be able to:

1. formulate a real problem in statistical terms;
2. give advice on efficient data collection;
3. analyse data and extract the maximum amount of information;
4. interpret and report the results.

In order to do this effectively, the statistician needs to know some background statistical theory and how to implement a range of standard statistical techniques (see Appendix A for a summary of what is required). However, statistics is much more than a collection of techniques and the statistician needs to be a **problem solver** and perhaps even a **policy-maker** (see Mosteller, 1988) rather than just a technician. Thus a statistician needs to understand the general principles involved in tackling statistical problems, and at some stage it is more important to study the **strategy** of problem solving rather than learn yet more techniques (which can always be looked up in a book). It is, for example, easy to **do** regression using modern computer software, but it is much harder to know when to use regression, which model to use, or even whether regression is appropriate at all.

The current literature concentrates heavily on techniques even though many topics currently taught are unlikely ever to be used by most statisticians, whereas they will, for example, find that it is commonplace to be given a set of messy data, with vague objectives, where it is unclear how to proceed. Such real-life problems generally fail to resemble the examples typically presented in textbooks. This book aims to redress the balance by presenting general advice on how to tackle real-life statistical problems. Of course general principles are difficult to specify exactly, and the advice in this book should be seen as guidelines rather than as a set of rigid rules.

Part Two of this book contains a series of worked exercises in question-and-answer format to give the reader some experience in handling real data. The importance of practical experience in developing statistical maturity cannot be over-stressed. Some may argue that limited teaching time is best spent on theory, and that the skills of a practising statistician are difficult to teach and only develop with genuine work experience in subsequent statistics employment. However, this attitude is unnecessarily defeatist. Some practical skills (e.g. report writing) can and should be taught formally, and there is increasing evidence that applied statistics can be taught effectively and that it provides a valuable complement to theory. Indeed, theory without practice is like a duck without water! Many colleges now offer courses involving a variety of practical work, ranging from straightforward exercises illustrating the use of techniques to substantial projects which mimic real-life situations so as to prepare students for the 'real world'.

So what do experienced statisticians actually do in practice? Is it what is in the textbooks? The simple answer is 'No'. Rather, the practising statistician needs to be versatile and resourceful, always being aware that the next problem may not fit any previously known recipe. Sound judgement is needed to clarify objectives, collect trustworthy data, and analyse and interpret them in a sensible way. It is particularly helpful to appreciate the different stages of a typical statistical investigation (see Chapter 2) from problem formulation right through to presenting the results, and to realize that the data analysis may take only a small fraction of the overall time and effort involved in any study.

It will become apparent that personal attributes such as an inquiring mind, an ability to communicate and good old-fashioned common sense (whatever that is!) are at least as important as knowing lots of statistical formulae (see the Summary at the end of Part One). Box (1976) says that the statistician's job requires among other things 'the wit to comprehend complicated scientific problems, the patience to listen, the penetration to ask the right questions, and the wisdom to see what is, and what is not, important'. Good statistical practice also requires the development of a number of personal skills which may not be covered in conventional statistical courses. They include the ability to write a good report (Chapter 11) and to use the computer and the library effectively (Chapters 8 and 9).

A number of themes will emerge during the course of the book. First, it is important to understand the **context** of any particular practical problem, and this means asking lots of questions to get sufficient background knowledge. This helps in formulating the problem, in building a sensible model, and in avoiding analysing every sample as if it were the only one ever taken.

A second theme is to avoid over-complicated and over-simple methods of analysis. Many people who use complex procedures have problems which should really be tackled in a simpler way. Expert advice on, say, factor analysis often convinces a client to adopt a simpler procedure. The use of unnecessarily complex methods means that attention is focused on the technical details rather than on potentially more important questions such as the quality of the data. At the other extreme, I have heard some practitioners say that we 'only need data collection and presentation', but this is being too simplistic. The emphasis in this book is towards the simpler end of the spectrum while hopefully avoiding the naive and simplistic.

There is particular emphasis on the **initial examination of data** (Chapter 6) as an essential precursor to a model-based analysis and as a tool for avoiding the over-complicated. While apparently simple, this is harder than it looks and is often skimped or otherwise done poorly. In particular, the presentation of graphs and tables is vital but often leaves much to be desired. The initial examination of data is vital, not only for summarizing data, but also for selecting and carrying out a 'proper' analysis correctly (not that there is anything improper about descriptive statistics!).

Finally, we note that a **robust near-optimal solution** may be preferred to an 'optimal' solution which depends on dubious assumptions. In particular, it is advisable to take active steps to avoid trouble at all stages of a statistical investigation. Textbooks are mostly concerned with 'optimal' solutions. With a single clearly defined objective (e.g. in a clinical trial) optimality considerations may indeed be paramount. But many (most?) studies are multi-purpose and it may not be possible to achieve simultaneous optimality. In any case optimality depends on **assumptions** which are unlikely to be satisfied exactly and may be seriously in error. Since many pitfalls await the unwary statistician, it is a good idea to look for a safe, robust, practical solution. Throughout the book, guidelines will be given for taking positive steps to avoid trouble and relevant examples include Examples 4.2–4.4, 6.2 and 10.2. Experience is a particularly good teacher for learning to sense where danger will be lurking (as in driving a car), but guidelines can still help.

Throughout the book (and especially in section 7.4), the reader is encouraged to take a flexible, pragmatic view of statistical inference. There are several different philosophical approaches to inference (e.g. frequentist and Bayesian), but it is argued that the statistician should adopt whichever approach is appropriate for a given problem rather than insist on using one particular approach every time.

1.3 Where does statistics stand today?

Statistical methods are widely used in all branches of scientific endeavour and methodological research continues to develop new and increasingly sophisticated techniques at an ever-growing pace. Yet my experience suggests that all is not well. There is a disturbing tendency for statistical techniques to be used by people who do not fully understand them. Non-statisticians seem to be getting more confused and the standard of statistical argument in scientific journals can best be described as variable. Complicated methods are often applied in a cookbook way which may turn out to be inappropriate or wrong, and authors often give lots of P-values rather than a clear description of their data. Thus statisticians should be concerned, not just with developing ever more complex techniques, but also with clarifying the general principles needed to apply the methods we already have. That is part of the motivation for this book.

For the future, I hope there will also be increased collaboration between statisticians and researchers in other disciplines, and I also expect much more research at the interface between statistics and computer science. Overall, I agree with Chambers (1993) that we need to take a broad view of statistics which is inclusive rather than exclusive, encourages cooperation with other sciences, is eclectic rather than dogmatic with regard to philosophy, and takes as much interest in problem formulation, data collection and the presentation of results as in data analysis.

1.4 Summary

To improve expertise in problem solving, the statistician needs more specific guidance on such topics as:

- problem formulation;
- model building;
- consulting;

- using the computer and the library;
- report writing.

Some recurring themes throughout the book are:

1. The need to understand strategy as well as know techniques.
2. The need to ask questions when formulating a problem and building a model so as to understand the practical context.
3. The importance of the initial data analysis, sometimes called exploratory data analysis, for checking data quality, summarizing data and helping to choose the appropriate follow-up analysis.
4. Avoiding trouble is complementary to, and a prerequisite for, finding an optimal solution.

Further reading

An excellent alternative discussion of some topics in Part One is given by Cox and Snell (1981, Part I). A detailed list of abilities (both technical and personal) required by a statistician is given by Anderson and Loynes (1987, section 2.2) together with suggestions as to how they might be taught. Some general advice on good statistical practice is given by Preece (1987). Statistical problem solving has much in common with problem solving in general and the mathematician's classic guide to problem solving by Pólya (1945) should be of interest. Some general advice on avoiding trouble is given by Chatfield (1991), and by the discussants of that paper, while Deming (1986) gives many cautionary tales from the business world.

2

The stages of a statistical investigation

A statistician should be involved at all the different stages of a statistical investigation. This includes planning the study, and then collecting, analysing, interpreting and presenting the results. This is much more satisfactory than being called in to analyse previously collected data or, even worse, being asked, for example, just to calculate a P-value.

When tackling statistical problems, it is helpful to bear in mind the different stages of an idealized statistical investigation which may be listed as follows:

1. Make sure you understand the problem and then formulate it in statistical terms. **Clarify the objectives** very carefully. **Ask as many questions as necessary.** Search the literature.
2. Plan the investigation and **collect** the data in an appropriate way (Chapter 4). It is important to achieve a fair balance between the effort expended in collecting the data and in analysing them. The method of collection is crucial to the ensuing analysis. For example the data from a proper randomized experimental design are quite different in kind to those resulting from a survey or observational study.
3. **Assess the structure and quality of the data.** The data will generally need to be **processed** onto a computer and this may involve **coding, typing** and **editing** (sections 6.2–6.4). Scrutinize the data for **errors, outliers** and **missing observations** and decide how to deal with any data peculiarities. For large data sets, the preliminary processing can take a high proportion of the time spent analysing the data.
4. Continue the **initial examination** of the data to obtain summary statistics, graphs and tables in order to **describe** the data and pick out interesting features of the data (section 6.5). The descriptive summary will sometimes be all that is necessary, but in other situations will be followed by more formal statistical inference as in stage 5 below.
5. Select and carry out an appropriate statistical analysis of the data (Chapter 7). The results of the descriptive analysis in stage 4 above will often be very helpful in formulating a model and choosing an appropriate follow-up procedure. Such procedures often assume a particular model structure, and

may involve estimating model parameters or testing hypotheses about the model. The fitted model needs to be checked, usually by looking at the residuals from the model (the differences between the observed and fitted values) to see if the model is adequate or if it needs to be refined or modified in some way.

6. Compare the findings with any previous results and acquire further data if necessary.

7. **Interpret and communicate the results** (Chapter 11). The findings may need to be understood by both statisticians and non-statisticians and this may entail separate reports. Extra care is needed in the presentation of graphs, tables and computer output.

This list is not sacrosanct and few investigations follow such a straightforward pattern in practice. In particular, there may be several cycles of model fitting as defects in some original model are recognized, further data are acquired and the model is gradually improved.

3
Formulating the problem

An approximate answer to the right question is worth a great deal more than a precise answer to the wrong question
– the first golden rule of applied mathematics

The formulation of a problem is often more essential than its solution which may be merely a matter of mathematical or experimental skill
– A. Einstein

The first step in any statistical investigation should be to:

- get a clear understanding of the physical background to the situation under study;
- clarify the objectives;
- formulate the problem in statistical terms.

Do not underestimate the time and effort required for this stage of the investigation.

If the investigation is to be carried out with other people in a consultative or collaborative project, the statistician must be prepared to **ask lots of questions**. We say more about this in Chapter 10. All variables which are thought to be important in the **context** of the particular situation need to be listed, and any distinctive features of the system, such as constraints and known limiting behaviour, should be noted.

It can be particularly difficult to specify the **objectives** clearly. They may range from a general desire for more information about, and increased understanding of, a particular problem, perhaps via an exploratory study, to a much more specific aim of testing a particular hypothesis, fitting a particular type of model, or choosing a course of action from a predetermined list of possibilities. It is always a good idea to think exactly how the results of a study will actually be used.

The objectives may sometimes be unclear even to the person who has asked the statistician for help, or may actually be quite different to those initially suggested (e.g. see the examples in Chapter 10 and Exercise B.3). Giving the 'right' answer to the wrong question is a more common error than might be expected and is sometimes called a **type III error** or an **error of the third kind** (Kimball, 1957). Thus finding the right question may be harder than finding the right answer.

Despite some effort, the objectives may sometimes remain rather ill defined. Moreover, the solution may turn out to be crucially dependent on whatever assumptions need to be made. Then there may not be a unique 'answer' but rather a range of possible solutions depending on a range of different assumptions. Statistical problems presented in textbooks usually have neat analytic solutions, whereas many real statistics problems do not. More guidance is needed on how to find good, but not necessarily optimal, solutions since near-optimality over a range of conditions is much more useful to the practitioner than exact optimality which is valid only under strict, possibly artificial, assumptions.

Other preliminary questions which need to be considered are the possible use of **prior information** and the **cost** and **likely benefits** of different strategies. Many projects are under-funded (as well as under-planned) and it may be necessary to say that the desired targets are unobtainable with the resources available.

A **literature search** is often advisable for revealing known facts, finding previous results for comparison, and perhaps even making further data collection unnecessary. The author was once asked to cooperate in conducting a sample survey to investigate a particular problem related to medical care, but half an hour in the library produced a published report on a previous survey which made the new survey redundant.

Even when a new investigation is thought necessary, it is still advisable to compare new results with previous ones so that established results can be generalized, modified or otherwise updated. There are many areas of science and social science where well-established laws already exist, as for example Boyle's law in physics and laws relating the market penetration and amount bought of different products in marketing research. Then the main question regarding new data is whether they agree with existing laws, and it is unfortunate that most statistics textbooks devote overwhelming attention to treating data as if they are always brand-new one-off samples. Further remarks on **model building** are made in section 5.3. A key question to ask regarding problem formulation is 'What is the model going to be used for?'. The answer should determine whether it is advisable, for example, to go for a simple or complicated model, and what level of descriptive accuracy is required.

Despite the importance of problem formulation, most teaching exercises concentrate on asking for solutions to ready-made, artificially clear-cut, questions. As a result, students get little practice at formulating problems and objectives and may end up learning the hard way, by making mistakes in real-life situations. This is a shame!

Further reading

Hand (1994) discusses problem formulation, which he calls 'deconstructing statistical questions'. Hamaker (1983) relates several instructive examples where he was asked to assist in designing an industrial experiment, but where an initial interrogation revealed that further background information was required, or even that no experiment was actually required.

4

Collecting the data

4.1 Methods of collecting data

In order to draw valid conclusions, it is important to collect 'good' data. It is usually helpful to distinguish between data collected in **designed experiments** (including **clinical trials**), **sample surveys**, pure **observational studies** and more specialist investigations such as controlled **prospective** and **retrospective** studies. The general principles of experimental design and survey sampling are summarized in sections A.10–A.12 of Appendix A but will probably be familiar to many readers. Here we make some more general comments.

Whatever method is used, the investigator needs to formulate objectives, decide which variables need to be measured and to what accuracy, and then specify the procedure for collecting data including such matters as sample size, the method of randomization to be applied, and how the data are to be recorded. Generally speaking, insufficient attention is given to the **management** of studies. Factors such as the order in which observations are taken, the effect of time of day, and the use of different observers can have substantial unforeseen effects. Moreover, while some plans can be specified exactly in advance, there will be other occasions when it becomes sensible to modify the procedures in an **iterative** way (Box, 1994) as exploration and discovery proceed. For example, a variable thought to be unimportant at the start may become a key variable as a result of early test results and require further detailed examination.

The **objectives** are usually crucial. The sort of design needed to pick out important variables (e.g. screening designs such as fractional factorials) are quite different to those concerned with modelling a response surface over a wide area or with choosing the values of explanatory variables so as to maximize a response variable. Sometimes the analyst is interested in modelling or maximizing a **mean** value of some response variable, while in other cases a reduction in **variability** is of prime interest. Sample surveys are usually concerned with collecting a variety of information and here the required precision may be critical.

The distinctions between the different types of data collection are not always clear-cut and there are many parallels between the basic concepts of design and survey sampling (e.g. compare blocking and stratification). However, there are substantial differences in spirit. Experiments require active intervention by the investigator, for example in deciding how to allocate treatments to experimental units. A clear interpretation of any differences which arise should then be possible. In contrast, as the name suggests, the investigator is generally more passive in observational studies and simply observes what is going on. Sample surveys involve drawing a representative sample from a well-defined population, usually by some sort of randomized procedure or by some sort of quota sampling. While well-conducted surveys make it possible to estimate population characteristics accurately, they are essentially observational in character and so suffer from the same drawbacks as observational studies, namely that it may be dangerous to try to interpret any interesting-looking effects which emerge, particularly as regards cause and effect. Thus although properly constructed observational studies can sometimes prove useful (e.g. Cochran, 1983), at least in providing useful pointers to check in subsequent studies, they can sometimes be worthless or even positively misleading.

Example 4.1 An observational study

I was given data for seven different treatments for child ear infections collected from all 400 + children who attended a local hospital over a six-month period. Large differences were observed between reinfection rates for the different treatments. These results provide a useful indication that some treatments are better than others. However, this was an observational study in which doctors had allocated the treatment they liked best, or thought was best for the patient. Thus the results need to be treated with caution. Some doctors always used the same treatment and then the effect of the treatment is **confounded with** (i.e. cannot be distinguished from) the effect of the doctor. Good results may indicate a good doctor, not a good treatment. Other doctors may have given particular treatments to particular groups of patients (e.g. the worst affected). Thus until a proper clinical trial is carried out, in which treatments are allocated **randomly** to patients, one cannot be sure that the observed differences between groups of patients getting the same treatment are actually due to the treatments. Nevertheless, historical data like these are better than nothing, and observational studies can be a useful cost-effective preliminary to a proper experiment.

Historical data are also used in controlled **retrospective** trials. Here a response variable is observed on carefully selected individuals and then the history of these individuals is extracted and examined in order to try and assess which variables are important in determining the condition of interest. Alternatively it is generally safer, but takes much longer, to assess important explanatory variables with a controlled **prospective** trial where individuals are chosen by the investigator and then followed through time to see what happens. The general term **longitudinal data** is used to describe data collected on the same units on several occasions over a period of time (e.g. see Plewis, 1985).

4.2 Choice of sample size

Whatever data-collection method is chosen, the investigator must select an appropriate sample size. This is a rather neglected topic, partly because it can be difficult to give general advice (and this section will be shorter than perhaps it should be!). When it is desired to test a particular hypothesis at a particular level of significance, the sample size can be chosen so as to achieve some required power for a particular alternative hypothesis. The level of significance is usually chosen to be 5%, while the power is often chosen to be 80% or 90%. Formulae are available to do this in standard situations, notably looking for a prespecified change in a mean value or in a proportion. However, in my experience these formulae are often used after the data have been collected in order to try and explain why a result was not significant!

In quality control and opinion sampling, sensible sample sizes are often well established from past experience. In other situations, investigators often take as large a sample as cost and time allow. With measured variables, a sample size of about 20 is usually a working minimum which can always be increased if necessary. With binary (success/failure) variables, much larger samples are needed. Thus to spot a 10% increase in the proportion of patients responding to a new treatment with a probability of, say, 80% (the power), a sample size of several hundred would be required.

Non-statisticians often choose a ridiculously small or unnecessarily large size. Many people think 'the larger the better', not realizing that a large, but poorly administered – and hence messy – sample may actually contain less information than a smaller, but carefully controlled, sample. On the other hand, some surveys are over-ambitious given the allowed sample size. For example, a study was suggested to compare several methods of treating elderly patients who have had a stroke and to use the results to compare four types of hospital, four types of specialty, five types of ward and to see if there is any variation during the year, all with a maximum of 240 patients. This size may sound quite large, but would give very small numbers in each cross-classified category.

4.3 Avoiding pitfalls when collecting data

The trials and tribulations involved in collecting data are best appreciated by actually doing it. Try selecting a 'random' live pig from a crowded pigsty and weighing it on a balance! Or try getting a quota sample to answer questions on some sensitive political topic. Refusal to cooperate and deliberate lies become a real problem. One interviewer confessed to me that the only way to get people to stop and answer questions was to stand next to a photo booth in a railway station, thus giving a biased sample even when quotas were satisfied. Generally speaking it is a good idea to carry out checks on the data collectors to see that they are doing their job properly.

Knowing the limitations of the data can be a big help in analysing them. To illustrate that a number may not be what it seems, I recall the case of a pregnant woman who went to hospital for a routine check, was kept waiting for three hours, was found to have abnormally high blood pressure, and so was admitted to hospital. Subsequent tests showed that she was actually quite normal and that the stress of waiting three hours had induced the abnormally high reading.

4.3.1 Sample surveys

There are many pitfalls in taking **sample surveys**. Some detailed advice is given in section A.10 of Appendix A. One important general point is that the **non-sampling errors** (such as non-response) are at least as important as **sampling errors** (due to looking at a sample rather than the complete population). The literature devotes much attention to finding 'optimal' sampling schemes so as to minimize the variance of a sample estimate subject to a cost constraint, but most of this theory ignores the multipurpose, multivariate nature of most surveys as well as disregarding non-sampling errors. The adjective **proximum** has been suggested for describing a design which in some sense aspires to approximate optimality while also taking account of practical factors such as robustness, administrative convenience and cost.

Two common failings are selecting an unrepresentative sample and asking silly questions. An unrepresentative sample may arise through using a judgement or convenience sample or through having an inadequate list (called a **frame**) of the population. Silly questions should be avoided by careful **questionnaire design**, which is very important both in deciding what questions to ask (do not try to ask too many) and how to word them. Questions must be fair, concise and clear. A **pilot survey** is vital for trying out a questionnaire.

The two other main sources of non-sampling error are **non-response**, due to unavailability of some sample members or a refusal to cooperate, and **measurement errors**, due for example to interviewer bias, recording errors and respondents giving incorrect information either deliberately or unwittingly.

Example 4.2 Some unrepresentative samples

1. When Puerto Rico was hit by a recent hurricane, there were 10 000 claims by residents for hurricane damage. The US government decided to base its total grant aid on finding the total of claims in the first 100 applications and then multiplying by 100. A colleague was involved in the difficult task of persuading the US government that the first 100 applications need not necessarily constitute a representative sample! Small claims are likely to come in first as they need less preparation. Fortunately the grant aid was eventually increased.

2. A company wanted to carry out a statistical analysis of a small 'random' sample of its clients. The selected sample comprised everyone whose surname began with the letter 'V'. This sample was selected by a management consultant with access to the database. The choice was made solely on the grounds of convenience, on the mistaken assumption that the alphabetical order of surnames is 'random' (which to some extent it is). However, the awful choice of the letter 'V' effectively excluded all Scottish and Welsh surnames, for example. Fortunately, the deficiency was easily spotted by the statistician who was asked to analyse the data.

3. I once analysed a large sample of observations (20 000 +) which I was told was a 1 in 20 systematic sample of the customers of a large national company. Although non-random, a systematic sample seemed reasonable here given that there is no apparent cyclic behaviour in the storage of customer details. Partly by good fortune, we realized that there was no one in the sample from a particular part of Britain and subsequent checks revealed that the data were stored on separate tapes (because of the large size of the data set) and that one tape had been missed out in producing the systematic sample. This sort of problem could become more widespread as large databases are increasingly under the control of some computing department and statisticians may not have direct access to the data.

4.3.2 Experimental designs

There are many possible failings in **experimental design**. Some general advice on 'good' design is given in section A.11 of Appendix A including a discussion of the principles of randomization, replication and blocking. Perhaps the most common failing is a lack of proper **randomization**, which can have an effect in all sorts of unforeseen ways. It is needed to eliminate the effect of nuisance factors and should be applied both to the allocation of experimental units and to the order in which observations are taken. For example, in regression experiments it is tempting to record the observations sequentially through time in the same order as the (increasing) values of the explanatory variable. Unfortunately, if the response variable also tends to increase, the researcher cannot tell if this is related to time or to the increase in the explanatory variable, or to both. The two effects cannot be disentangled and are said to be **confounded**. If randomization is left to a 'client' it is often carried out incorrectly, and so the document specifying how to carry out the design (called the **protocol**) should spell out exactly how it should be done.

Check that the people carrying out the experiment know exactly what to do (Example 4.4), and that the design takes account of the particular practical situation (Example 4.3). There is no point in devising a brilliant plan if it is wrongly implemented or ignores some local feature. For example, every field plan for an agricultural trial should have a compass indication as to where North is!

Sometimes a pilot study can also be useful in experimental design to determine background variability and assess ways of actually collecting the observations. This may reveal unforeseen problems and lead to the protocol being redesigned.

On sample size, check that there are enough degrees of freedom to estimate the residual variance adequately. The dangers of saturated designs in Taguchi methods are highlighted by Bissell (1989).

Another general piece of advice is to ensure that replicates really are independent, and try to avoid what Hurlbert (1984) calls **pseudo-replication** (Example 4.5).

Finally, it is worth noting that there is a vast difference between the sort of design suitable for testing a single new treatment and the sort of screening design appropriate in an agricultural setting for such products as insecticides, fungicides, herbicides and fertilizers. An agricultural products research station may carry out primary screening of as many as 10 000 new products a year. Of these maybe 1000 will proceed to secondary testing and 100 to tertiary testing in proper field trials. After this there will still be statutory tests for efficacy (does it work?), for toxicology (is it harmful?), and for environmental soundness (does it damage the ecosystem?). The actual number of products launched will be in single figures or even zero!

Clinical trials (see section A.12 of Appendix A) are particularly prone to error and Andersen (1990) devotes an entire book to an alarmingly diverse collection of flawed experiments. The basic problem is that two types of people are involved in drug trials on humans, namely doctors and patients, and both groups can act perversely in such a way as to ruin the design. For example, a doctor may choose to ignore the protocol and allocate a treatment which he believes is best for the patient, while a patient may not take the drug as directed.

Example 4.3 Some faulty experimental designs

1. Preece (1987) recounts an experience visiting a forestry experiment on some sloping land in Africa. When he commented that the design did not appear to match up with the terrain, he was told that the experiment had been designed in Rome! While it may be reasonable to select the treatments in Rome, and perhaps the size and type of design, the details of the field plan should have been resolved on the spot.
2. In a particular field experiment, the treatments had to be applied on two separate occasions, but the field worker read the plan from the wrong end on the second occasion. The treatment combinations were therefore quite different from what had been planned.
3. In a factorial design to assess the effect of a fungicide on planks of wood, four factors were considered, namely (i) treated/untreated; (ii) side or ends of planks; (iii) Mississippi or Wisconsin; (iv) month of year. Counts were taken of the number of fungi on each plank after an appropriate period.

Surprisingly, a large treatment effect was found in Wisconsin but not in Mississippi. How could this be? Further investigation revealed that in Mississippi the planks treated with fungicide had been placed on top of the untreated planks. Presumably the fungicide had run in the rain and treated the planks lower down. An elementary error perhaps, but an example of the sort of disastrous incident which can ruin the best-laid plans.

Example 4.4 A suspect recording procedure

This example illustrates the difficulties which can arise in recording measurements in real life. A scientist wanted a function to predict the conductivity of a copper sulphate solution. An experiment was carefully designed (actually a type of composite factorial experiment) with three control variables, namely $CuSO_4$ concentration, H_2SO_4 concentration, and temperature. The results were collected and a full quadratic model was fitted using least-squares regression analysis. An almost perfect fit was expected, but in fact only about 10% of the variation was explained by the fitted model. The scientist then complained to a statistician that there must be something wrong with the regression package.

The statistician questioned the scientist carefully and found that the observations had been collected, not by the scientist himself, but by a technician who was thought to be reliable. The experiment consisted of making up solutions in beakers which were put onto temperature control pads. When they reached the design temperature, a conductivity probe was held in each beaker and a reading was taken from a digital panel. The statistician asked for the experiment to be repeated so that he could see exactly how the observations were taken. Two problems were soon spotted. Sometimes the probe (held by the technician) touched the side or bottom of the beaker and sometimes it was not immersed deeply enough. Furthermore, the digital readout took about 15 seconds to settle to a stable value (the probe required diffusion of the solution through a membrane), but the technician often did not wait that long. The statistician therefore made two practical recommendations. First, the probe should be fixed in a clamp so that it always went to the same depth. Second, a timer should be used so that the conductivity was not read until 30 seconds after immersion. The whole experiment was repeated in less than one hour, the regression model was refitted and this time an excellent fit resulted ($R^2 = 0.99$). The scientist was delighted.

An obvious question is whether the two recommendations made above are 'statistical'. At first sight they may appear to be technological, rather than statistical. However, I would take a wider view of the statistician's responsibilities as including good scientific method in general. We should not see ourselves as technical assistants to scientists or as second-class mathematicians, but rather we should 'strive to earn the title of first class scientist' (G.E.P. Box).

Example 4.5 Pseudo-replication

Suppose you want to determine how leaves from a certain type of tree decompose on a lake bottom at a depth of 1 metre. You fill eight small net-bags with leaves and place them at the same location. After one month you retrieve the bags and measure decomposition. You have eight replicates, or do you? In fact you have no information about how decomposition varies from one spot to another at 1 m depth. The observations are not independent and no generalization is possible.

The problem is even harder to spot when comparing eight bags at one location at one depth with eight bags at a different location at a different depth. Does a significant difference indicate a difference between the effect of different depths or is it just evidence of a difference between the two locations? Put another way, what is the 'treatment'?

Example 4.6 Collecting 'official' statistics

This final cautionary tale should provoke a healthy distrust of official recorded data and emphasize that it is not enough to collect lots of data. A prominent British politician (Healey, 1980) recounts how he was appointed 'railway checker' at a large railway station during the Second World War. He was expected to count the number of servicemen and servicewomen getting on and off trains, but as a convalescent with eight platforms to cover in blackout conditions, he made up all the figures! He later salved his conscience by asking ticket collectors to provide numbers leaving each train, but subsequently discovered that these number were invented as well! Fortunately the war effort did not suffer visibly as a result!

The obvious moral of this story is that there must be a sensible balance between effort spent collecting data and analysing them, and that care is needed to specify exactly what information is worth recording and how. At the time of writing (1994) there has been an explosion in the amount of data that hospital managers are trying to collect in the UK so that nurses and doctors spend a high proportion of their time completing poorly designed forms rather than treating patients. Much of the information is of dubious quality and some is never actually used. Insufficient attention has been given to specifying objectives and to questionnaire design.

4.4 Summary

For sample surveys, check that:

- the sample is representative and of a suitable size;
- questions are fair, clear and concise (they should be tested in a pilot survey);
- non-sampling errors, such as interviewer bias, non-response and measurement error, are reduced as far as possible;
- data are coded and processed accurately.

For experimental designs, check that:

- randomization really is used properly;
- it is possible to find estimates of all effects of interest, including potentially important interactions, and give them a unique interpretation (which is not possible in Examples 4.1 and 4.5 above).

Overall, it is clear that a statistician who is involved at the start of an investigation, who advises on data collection, and who knows the background and objectives, will generally be able to make a much more satisfying contribution to a study than a statistician who is called in later on. Unfortunately, data are still often collected in a rather haphazard way without the advice of a statistician. When this happens, the statistician who is asked to analyse the data should closely scrutinize the data-collection procedure to find out exactly how they were collected. This will determine whether it is worth spending much effort on the analysis. The key messages are the following:

1. Data collection is as much a part of statistics as data analysis.
2. Do not trust any data that you did not collect yourself.

Further reading

Some general references on sample surveys, experimental design and clinical trials are given in sections A.10, A.11 and A.12, respectively, of Appendix A.

5
Analysing the data — 1: General strategy

Having collected (or been given) a set of data, the apprentice statistician may be bewildered by the wide variety of statistical methods which are available and so be unsure how to proceed. The details of standard techniques are well covered in other books and will not be repeated here, although a concise digest of selected topics is given in Appendix A. Rather this book is concerned with general **strategy** such as how to:

- process the data;
- formulate a sensible model;
- choose an appropriate method of analysis.

5.1 The phases of an analysis

Students are often given the impression that a statistical analysis consists of 'doing a t-test' or 'fitting a regression curve'. However, life is rarely as simple as that. It is helpful to distinguish five main stages of a 'typical' analysis as follows:

1. **Look at the data.** Summarize them using simple descriptive techniques. Evaluate the quality of the data and modify as necessary.
2. **Formulate a sensible model.** Use the results of the descriptive analysis and any other background information.
3. **Fit the model to the data.** Estimate the model parameters. One or more hypotheses may be tested at this stage. Ensure that the fitted model reflects the observed systematic variation. Furthermore, the random component needs to be estimated so that the precision of the fitted model may be assessed.
4. **Check the fit of the model.** The assumptions implicit in the fitted model need to be checked. It is also a good idea to see if the fit is unduly sensitive to a small number of observations. Be prepared to modify the model if necessary.

5. **Utilize the model and present the conclusions**. The model may be used for descriptive, predictive or comparative purposes. A summary of the data, the fitted model and its implications need to be communicated. This should include a statement of the conditions under which the model is thought to be applicable and where systematic deviations can be expected.

It is sometimes helpful to regard these five stages as forming two main phases which can be described as:

- the **preliminary** analysis;
- the **definitive** analysis.

One objective of this book is to emphasize the importance of the preliminary phase, which does not always receive the attention it deserves. It is essentially stage 1 above and, in more detail, includes:

- Processing the data into a suitable form for analysis. This usually includes getting the data onto a computer.
- Checking the quality of the data. It is important to recognize the strengths and limitations of the data. Are there any errors or wild observations (called **outliers**)? Are there missing observations? Have the data been recorded to sufficient accuracy?
- Modifying the data if necessary, for example, by transforming one or more variables and by correcting errors.
- Summarizing the data using simple descriptive statistics including graphs and tables.

The definitive analysis is often based on a probability model and may involve parameter estimation and hypothesis testing. However, it is a common mistake to think that the main analysis consists only of model fitting. Rather it includes all three stages of model building, namely model formulation, fitting and checking. Thus the definitive analysis includes stages 2, 3 and 4 above, and may also be taken to include the presentation of the conclusions in a clear and concise way. Of course, if a well-established model already exists, based on many previous data sets, then the definitive analysis is concerned with assessing whether the new data conform to the existing model rather than with fitting a new model from scratch. Whatever the situation, one important overall message is that the analyst should not be tempted to rush into using a standard statistical technique without first having a careful look at the data. This is always a danger now that easy-to-use computer software is so widely available.

In practice, the two phases (and the five stages) generally overlap or there may be several cycles of model fitting as the model is gradually improved, especially if new data become available. Thus the distinction between the two phases, while useful, should not be taken too seriously. In particular, the preliminary analysis may give such clear-cut results that no follow-up definitive analysis is required. For this reason, I prefer to use the neutral title 'initial examination of data', rather than 'preliminary analysis', although it is in the same spirit.

The initial examination of data will be discussed more fully in Chapter 6, while the choice of definitive analysis is discussed in Chapter 7. The remainder of Chapter 5 is mainly concerned with general aspects of model building.

5.2 Other aspects of strategy

It is sometimes helpful to distinguish between methods of analysis which do or do not depend on a formal probability model. The former can be called **probabilistic**, while the latter are called **descriptive** or **data-analytic**. For example, simple graphical methods are descriptive while regression analysis is probabilistic. Of course, many statistical problems require a combination of both approaches and there is in any case some overlap. For example, by quoting the sample mean, rather than the sample median, in a data summary, the analyst implicitly assumes that a sensible model for the underlying distribution of the observations is approximately symmetric.

Another distinction that is sometimes made is that between an **exploratory** and a **confirmatory** analysis. The former is supposed to be concerned with a completely new set of data where there is little or no prior knowledge about the problem. In contrast, a confirmatory analysis is used to check the presence or absence of a phenomenon observed in a previous set of data or expected on theoretical grounds. In practice, pure exploratory studies should be rather rare as there is nearly always some prior information; pure confirmatory analyses also seem rather rare, except in drug testing. Perhaps a better distinction is that between exploratory and confirmatory methods, since exploratory methods, for example, are not necessarily concerned with a completely new set of data. Rather most statistical investigations involve elements of both types of method (Cox, 1990). For example, it is wise to carry out an initial examination of the data, using simple descriptive methods, even when the emphasis of the work is confirmatory.

A large proportion of 'significant' results reported in the literature, particularly in medicine and the social sciences, are essentially of one-off exploratory data sets, even though it is known to be dangerous to spot an 'interesting' feature on a new set of data and then test this feature on the same data set (section 7.2). Here confirmatory methods are being used, rather inappropriately, on exploratory data. Generally speaking, there is too much emphasis in statistical work on analysing single data sets in isolation (called the 'cult of the isolated study' by Nelder, 1986).

In contrast, there is not enough help in the literature on how to combine information from more than one experiment. The rather horrid term **meta-analysis** (e.g. see the review articles in *Statistical Science*, 1992, **7**, Part 2) has recently been coined to describe the combination of information from two or more studies. Originally this was primarily concerned with finding an overall *P*-value for several studies comparing two treatments. However, a broader view should now be taken of methods for combining information from replicated research studies, since this is a fundamental scientific activity. For example, a summary of six studies might conclude that two studies should be disregarded because the designs were flawed while the other four studies all point to a substantial difference between the effects of two treatments. This difference may then be quantified and related to the perceived quality of the different trials and to the conditions under which they were run.

Of course an accepted scientific relationship (such as Boyle's law relating the pressure and volume of a gas) is based not on a single 'best-fit' analysis but on numerous studies carried out under a variety of conditions, and it is not clear if this should be regarded as a meta-analysis. We really need to develop methods for analysing a series of similar or related data sets, such as those arising in marketing research from ongoing surveys of consumer purchasing behaviour, where descriptive comparisons of the different data sets allow one gradually to build up a comprehensive, empirically-based model which can be used with some confidence. Some probabilistic mechanism may also be found to explain what is going on. This approach is amplified in sections 5.3.3 and 7.4.

5.3 Model building

All models are wrong but some are useful – *G.E.P. Box*

In most statistical investigations, it is helpful to construct a **model** which provides a mathematical representation of the given physical situation. This model should provide an adequate description of the given data and enable predictions and other inferences to be made. The model may involve unknown coefficients, called **parameters**, which have to be estimated from the data.

There are many different types of model. Statistical models usually contain one or more systematic components as well as a random (or stochastic) component. The **random** component, sometimes called the **noise**, arises for a variety of reasons and it is sometimes helpful to distinguish between (i) measurement error and (ii) natural random variability arising from differences between experimental units and from changes in experimental conditions which cannot be controlled. The **systematic** component, sometimes called the **signal**, may be deterministic, but there is increasing interest in the case where the signal evolves through time according to probabilistic laws. In engineering parlance, a statistical analysis can be regarded as extracting information about the signal in the presence of noise. Some models are tailored to the specific subject-matter situation (called **substantive** models by Cox (1990)). An example might be a stochastic process model for the spread of AIDS. A model, such as a regression or ANOVA model (described by Cox (1990) as an **empirical** model), aims to capture some sort of smooth average behaviour in the long run. Such models are not based on highly specific subject-matter considerations and this can be seen as a strength (they are widely applicable) or as a limitation. In contrast to (signal + noise) models, many useful scientific models are of a completely different type in that they make no attempt to model the process mechanism or even to model measurement errors. They are essentially **descriptive** and usually empirically-based. An example is Boyle's law in physics which says that pressure × volume ≃ constant for a given quantity of gas, when external conditions are kept fixed. Notice the absence of any attempt to model variability.

Whatever sort of model is fitted, it should be remembered that it is impossible to represent a real-world system exactly by a simple mathematical model. However, it is possible that a carefully constructed model can provide a good approximation, both to the systematic variation and to the scatter. The challenge for the model builder is to get the most out of the modelling process by choosing a model of the right form and complexity so as to describe those aspects of the system which are perceived as important.

As noted in Chapter 3, a key question concerns what the model is going to be used for. There are various objectives in model building (see, for example, Daniel and Wood, 1980; Gilchrist, 1984):

1. To provide a parsimonious summary or description of one or more sets of data. By **parsimonious**, we mean that the model should be as simple as possible (and contain as few parameters as possible), consistent with describing the important features of the data.
2. To provide a basis for comparing several different sets of data.
3. To confirm or refute a theoretical relationship suggested a priori.
4. To describe the properties of the random or residual variation, often called the **error component**. This will enable the analyst to make **inferences** from a sample to the corresponding population, to assess the precision of parameter estimates, and to **assess the uncertainty** in any conclusions.
5. To provide **predictions** which act as a 'yardstick' or norm, even when the model is known not to hold for some reason.
6. To provide **physical insight** into the underlying physical process. In particular, the model can be used to see how perturbing the model structure or the model parameters will affect the behaviour of the process. This is sometimes done analytically and sometimes using simulation.

You should notice that the above list does not include getting the best fit to the observed data (see the ballade of multiple regression in section A.6 of Appendix A). The term **data mining** (e.g. Lovell, 1983) is sometimes used to describe the dubious procedure of trying lots of different models until a good-looking fit is obtained. However, the purpose of model building is not just to get the 'best' fit, but rather to construct a model which is consistent, not only with the data, but also with background knowledge and with any earlier data sets. In particular, the choice between two competing models which fit a set of data about equally well should be made on grounds external to the data or on the model's predictive ability with new data. The ability of a model to give a good fit to a single set of data may indicate little since a good fit can always be achieved by making the model more and more complicated. This raises the general problem of **overfitting**. Fitting an over-complicated model may give a good fit and give predictions with low bias but a high prediction error variance may result. In contrast, underfitting will generally give lower variance but higher bias. As always, a compromise must be made.

Some statistical procedures place great emphasis on making a careful study of the process by which data are generated, and then constructing a parsimonious model of it. However, there are a number of techniques, often called 'black-box' techniques, which avoid making substantive parametric assumptions and tackle problems in a way which is often computationally intensive. Here the 'model' has much lower importance. Both types of approach have their place for different types of practical problem, depending partly on the amount of prior information available, and the use, if any, to be made of the model.

Occasionally it may be useful to employ more than one model. For example, in long-range forecasting it is often advisable to construct a range of forecasts based on a variety of plausible assumptions about the 'true' model and about what the future holds. This is sometimes called **scenario forecasting**.

There are three main stages in model building, when starting more or less from scratch, namely:

1. model formulation (or model specification);
2. estimation (or model fitting);
3. model validation.

In introductory statistics courses there is usually emphasis on stage 2 and, to a lesser extent, on stage 3, while stage 1 is often largely ignored. This is unfortunate because model formulation is often the most important and difficult stage of the analysis, while estimation is relatively straightforward, with a well-developed theory, much easy-to-use computer software and many reference books. This book therefore concentrates on stages 1 and 3.

The three stages are discussed in turn, although in practice they may overlap or there may be several cycles of model fitting as the model is refined in response to diagnostic checks or to new data. Model building is an iterative, interactive process (Box, 1976; 1980; 1994).

5.3.1 Model formulation

The general principles and skills involved in model formulation are covered in some books on scientific method and on mathematical modelling (e.g. Edwards and Hamson, 1989) but are rather neglected in statistics textbooks. The analyst should:

1. Consult, collaborate and discuss with appropriate experts on the given topic. **Ask lots of questions** (and listen). The **context** of any problem is crucial and you should not try to model without understanding the subject-matter aspects of the system under study; these include commercial constraints, background theory (see below) and the desirability of keeping the model simple.
2. **Incorporate background theory**. This should suggest which variables to include, and in what form. Sometimes external considerations indicate that a variable should be included even when the analysis appears to suggest it need not be. You should also take account of known constraints on the variables, for example that a relationship should be monotonically increasing, or that a coefficient should be non-negative. Known limiting behaviour should also be incorporated such as behaviour at infinity or zero; there is, for example, a crucial difference between a function which increases without limit and one which tends in the limit to a finite value (e.g. Edwards and Hamson, 1989, section 5.4).
3. **Look at the data** and assess their more important features; see the remarks in Chapter 6 on the initial examination of data.
4. Incorporate information from other similar data sets (e.g. any previously fitted model).
5. Check that a model formulated on empirical and/or theoretical grounds is consistent with any qualitative knowledge of the system and is also capable of reproducing the main characteristics of the data.
6. Remember that all models are **approximate** and **tentative**, at least to start with; be prepared to modify a model during the analysis or as further data are collected and examined.
7. Realize that **experience** and **inspiration** are important and that many subjective choices have to be made.

It is clear that model formulation is a complex matter which is strongly influenced by **contextual considerations**, by the **intended use** of the model, as well as by the prior perspective regarding the class of models to be entertained. Although a fundamental data-analytic task, there is no widely accepted general strategy for building models and much of the available theory is concerned with the much more limited task of distinguishing between two or more prespecified models.

At all stages of model formulation, it is helpful to distinguish between

- what is known with near certainty;
- what is assumed on some reasonable basis;
- what is assumed for mathematical convenience;
- what is unclear.

All modelling involves **assumptions** of varying credibility. For example, 'a man in daily muddy contact with field experiments could not be expected to have much faith in any direct assumption of **independently** distributed normal errors' (Box, 1976). Too often we make such assumptions (especially about the 'errors') without giving them much thought.

It is worth noting that the extent to which model structure should be based on background theory and/or on observed data is the subject of some controversy. For example, in time-series analysis an econometrician will tend to rely more on economic theory while a statistician will tend to rely more on the properties of the data. To some extent this reflects different objectives, but it also indicates that model building depends partly on the knowledge and prejudices of the analyst. While some analysts wrongly ignore theoretical reasoning, others place too much faith in it and it is obviously important to let the data speak for themselves when there are doubts about theory or when theories actually conflict. As always, a combination of theory and data analysis is fruitful.

In constructing a model, it is helpful to distinguish between various aspects of the model. First, many models contain separate **systematic** and **random** components. The specification of the random component will typically include the form of the distribution as well as whether successive values are independent with constant variance. The distribution is often called the 'error' distribution, though the use of the word 'error' in this context does not imply a mistake, and many authors prefer an alternative term such as 'deviation'. A second useful distinction is that between the **primary** assumptions, judged central to the problem, and the **secondary** assumptions, which are less important and where some approximation/simplification can often be made. For example, in regression analysis the form of the curve (linear, quadratic or whatever) is usually the primary focus of interest while the 'error' assumptions are usually secondary (though this is not always the case). A third useful distinction is that between the model parameters of direct interest and the parameters, called **nuisance** parameters, which are not of direct interest. For example, the 'error' variance is often a nuisance parameter.

Example 5.1 A regression model

Consider the problem of formulating a regression model relating observations on a response variable, y, to an explanatory variable x. The regression curve constitutes the systematic component while the deviations from the curve constitute the random part. As noted above, the form of the regression curve normally constitutes the primary assumptions and there may be background theory or previous data modelling to suggest a particular form of curve or at least to put some restrictions on it (e.g. y may be constrained to be non-negative). In addition, the analyst should look at a scatter plot of the data to assess the form of the relationship. As regards the random component, the analyst must use past data and an assessment of the scatter plot to make assumptions about the conditional distribution of y for a fixed value of x. Is the distribution (approximately) normal? Is the conditional variance of y constant or does it vary with x? Are the deviations independent? The standard regression model assumes that 'errors' are independent and normally distributed with zero mean and constant variance. These are a lot of assumptions to make and they need to be checked, usually at the model-validation stage (see section 5.3.3).

The effects of model uncertainty

Nowadays computers let us look at tens or even hundreds of models at the model-formulation stage. As well as trying different forms for the systematic component, the analyst may try including/excluding different variables, transforming one or more variables, omitting or adjusting outliers, and so on. A case study with realistically messy data illustrating the difficulties of fitting a multiple regression model to data with as many as 20 explanatory variables is given by Janson (1988).

The facility to try many models is very helpful in many ways, but can also be dangerous. It can result in overfitting and in underestimating the precision of model predictions. By trying so many models, the analyst implicitly admits that the form of the model is uncertain. Yet, having selected a model, often using some sort of 'best-fit' criterion, the analyst generally behaves as if the model is known to be true. Then the estimates of precision which result (e.g. the width of confidence intervals) will typically take account of sampling variability but not of model uncertainty. Unfortunately, if a model is formulated from, and then fitted to, the same set of data, it is known that least-squares theory no longer applies. Large biases have been demonstrated, for example when subset selection methods are used in multiple regression (see Miller, 1990), and in time-series analysis when a best-fitting model is chosen from a wide family of models (Chatfield, 1995). Likewise in hypothesis testing, if a hypothesis is generated and then tested on the same data, then the P-value can be misleading (section 7.2). These difficulties, essentially due to model uncertainty, are largely ignored by statisticians partly because no simple solutions have yet been found when using a frequentist approach, though some limited progress can be made using simulation and resampling methods.

An alternative approach using Bayesian methodology has been reviewed by Draper (1995). This does not require the analyst to select a single 'best' model, but rather averages over a number of competing models. If prior probabilities can be attached to the different plausible models, then this approach will give a more realistic assessment of the effect of model uncertainty.

Is there a 'true' model?

Another crucial question in model building is whether one believes there really is a 'true' model. Most statisticians do not, but rather see models as (hopefully) useful approximations. Hence the quote by George Box at the start of this section. The absence of a true model makes the arguments about different approaches to statistical inference (section 7.4) seem rather academic. Model builders increasingly adopt a pragmatic approach which concentrates on finding a parsimonious description of the data and then assessing its accuracy and usefulness, rather than just testing it. The context and the objectives (rather than just finding the 'best fit') are key factors in deciding whether a model is useful.

5.3.2 Model estimation

The model-fitting stage consists primarily of finding point and interval estimates of the model parameters. Section 7.4 discusses the different philosophical approaches to inference, while Appendix A.4 reviews the technical details of estimation including terminology, some general point estimation methods, robust estimation and bootstrapping. Some tests of hypotheses regarding the values of the parameters may also be carried out at this stage (section 7.2 and Appendix A.5).

The wide availability of computer packages makes it relatively easy to fit most standard models. It is worth finding out what estimation procedure is actually used by a given package, assuming that the information is available in the documentation (as it should be). In addition to point estimates, a good package should also provide standard errors of estimates as well as various derived quantities such as fitted values and residuals which will in turn help in model validation. No further details will be given here on what is usually the most straightforward part of model building.

5.3.3 Model validation

When a model has been fitted to a set of data, the underlying assumptions need to be checked. If necessary, the model may need to be modified. Answers are sought to such questions as:

1. Is the systematic part of the model satisfactory? If not, should the form of the model be altered, should some variables be transformed, or should some additional variable(s) be included? Can the model be simplified in any way, for example by removing one or more variables?
2. Is the random component of the model satisfactory? What is the distribution of the 'error' terms? (Many models assume approximate normality.) Is the

error variance really (more or less) constant (as assumed by many models and by many model-fitting procedures)? If not, can the model-fitting be suitably adapted, for example by using **weighted least squares**?

3. How much does the 'good fit' of the model depend on a few **influential** observations?
4. Has any important feature of the data been overlooked?
5. Are there alternative models that fit the data nearly as well but lead to substantially different predictions and conclusions? If so, the choice of model should be based on subject-matter considerations or on predictive ability. Occasionally a larger combined model should be considered or a range of predictions based on different models.
6. Be prepared to **iterate** towards a satisfactory model.

Model validation is variously called **model checking, model evaluation, diagnostic checking** or even **residual analysis** as most such procedures involve looking at the residuals. A **residual** is the difference between an observed value and the corresponding fitted value found from the model. This can be expressed in the important symbolic formula

$$DATA = FIT + RESIDUAL$$

For some purposes, it is preferable to convert these raw residuals into what are called **standardized residuals** which are designed to have equal variance. This can readily be done by a computer. Some technical details on this and other matters are given in Appendix A.8. There are many ways of examining the residuals, and the choice depends to some extent on the type of model. For example, a good computer package will plot the values of the residuals against the values of other measured variables to see if there is any systematic trend or pattern. An example of such a graph is given in Exercise C.4. It is also advisable to plot the residuals in the order in which they were collected to see if there is any trend with time. The residuals should also be plotted against the fitted values and against any other variable of interest. If the residuals reveal an unexpected pattern, then the model needs appropriate modification.

The distribution of the residuals should also be examined, perhaps by simply plotting their histogram and examining its shape, or by carrying out some form of probability plotting (Appendix A.1). However, note that the distribution of the residuals is not exactly the same as the underlying 'error' distribution, particularly for small samples. Various tests for normality are available (e.g. Wetherill, 1986, Chapter 8) but I have rarely needed to use them in practice because the analyst is usually only concerned about gross departures from assumptions which are usually obvious 'by eye'.

There is usually special interest in large residuals. A large residual may arise because:

1. the corresponding observation is an error of some sort;

2. the corresponding observation is a genuine, but extreme, value, perhaps because the 'error' distribution is not normal but skewed so that occasional large residuals are bound to arise;
3. the wrong model has been fitted or an inappropriate form of analysis has been used.

In practice, it is often difficult to decide which is the correct explanation. In particular it is potentially dangerous to assume 1 above, and then omit the observation as an outlier, when 2 or 3 is actually true. Further remarks on dealing with outliers are given in section 6.4.3.

It is also useful to understand what is meant by an **influential** observation, namely an observation whose removal would lead to substantial changes in the fitted model (Appendix A.6 and A.8). A gross outlier is usually influential, but there are other possibilities and it is wise to find out why and how an observation is influential. It is certainly unwise to throw out observations just because they are unusual in some way.

It is also important to distinguish between gross violations of the model assumptions and minor departures which may be inconsequential. A statistical procedure is said to be **robust** if it is not much affected by minor departures, and it is fortunate that many procedures have this property. For example, the t-test is robust to moderate departures from normality.

Finally, we note a point which is widely ignored in the statistical literature, namely that diagnostic checks on a single data set have rather limited value and can be overdone. There are statistical problems in constructing and validating a model on the same data set (see also sections 5.3.1 and 7.2) since the tests typically assume the model is specified a priori and calculate a P-value as:

Probability (more extreme result than the one obtained | model is true)

However what is really required is:

Probability (more extreme result than the one obtained | model has been selected as 'best' by the model-formulation procedure)

It is not clear in general how this can be calculated.

What is clear is that the good fit of a 'best-fitting' model should not be surprising. Moreover, such checks of a single data set ignore the deeper issue as to whether the results generalize to other data sets and to other observational conditions. If they do not, then the occurrence in the first data set was merely a one-off happening of no general interest. Science is concerned with getting **repeatable** results (or at least it should be). It is most unfortunate not only that statistics concentrates on single sets of data, but also that checking repeatability is now regarded as a 'waste of time' in some so-called scientific areas (see Feynman, 1985, pp. 338–346.) Thus I would like to see model validation expanded to include checking a model on further data. In principle this is usually possible and in practice should be pursued with more determination. For example, I have worked for a number of years on a model to describe the purchasing behaviour of consumers of various manufactured products, such as soap and toothpaste (e.g. Goodhardt, Ehrenberg and Chatfield, 1984). The model has been found to 'work' with data collected at different times in different countries for different products. This sort of model is far more useful than, say, a regression model which typically just happens to give the best fit for one particular set of data.

Thus methods for analysing models with more than one set of data deserve more attention in the statistical literature – and I am not thinking here of meta-analysis (section 5.2). More guidance is particularly needed on **tuning** and **extending** models as more data become available, perhaps collected under somewhat different conditions. Model tuning involves making (usually minor) adjustments to a model, for example by modifying the estimate of a model parameter. Model extension involves expanding a model, for example by extending the range of conditions under which is is known to be valid.

5.3.4 Utilizing a model

Once a model has been fitted and checked, it may be used for whatever purpose it has been built. This may include **describing** data, **making predictions** and enabling **comparisons** to be made with the results of similar data sets. As always, the analyst should have clear objectives and know what problem has to be solved.

As regards making predictions, it is worth noting that statistical inference concentrates on estimating **unobservable** quantities, such as model parameters. This is of course important for describing and understanding the given situation, but one rarely knows if such inferences are 'good' since the estimates cannot be compared directly with the truth. In contrast, it is comparatively easy to check whether predictions of **observable** quantities (e.g. Geisser, 1993) are 'good' and the value of a model can be judged by assessing and **calibrating** such out-of-sample predictions, especially when the predictions are made under the range of conditions for which the model is known to hold. An example is using a regression model to make predictions of future values within the range of previously explored values of the explanatory variable(s).

In practice a model is often used to make predictions outside the established range of conditions for which the model is known to be valid. Such **extrapolation** is well known to be dangerous, as illustrated by the US Space Shuttle catastrophe in 1986 (Chatfield, 1991, Example 8). Extrapolations rely on the in-built model assumptions being true outside previously studied conditions. While dangerous, this is sometimes unavoidable, as in time-series forecasting where forecasts of future values are required. Many business and economic forecasts have gone wrong in the past (who could have predicted the Gulf War or the sudden removal of the 'iron curtain'?) and the cautious analyst will not only give a prediction interval, rather than a point forecast, but also increase the width of the prediction interval to allow for model uncertainty or give a range of predictions based on different, but clearly stated, assumptions (or scenarios).

Further reading

Model formulation is discussed by Cox and Snell (1981, Chapter 4) and in much more detail by Gilchrist (1984). Some further general philosophical remarks on modelling are given in section 7.4.

6
Analysing the data — 2: The initial examination of data

6.1 Introduction

It is usually best to begin an analysis with an informal exploratory look at a given set of data in order to get a 'feel' for them. As they gain more experience, statisticians typically find that they devote more and more energy to this important first phase of the analysis (section 5.1) which comprises the following steps:

1. Process the data into a suitable form for analysis.
2. Check the quality of the data. Are there errors, outliers, missing observations or other peculiarities? Do the data need to be modified in any way?
3. Calculate simple descriptive statistics. These include summary statistics and graphs.

The general aim is to summarize the data, iron out any peculiarities and perhaps get ideas for a more sophisticated analysis. The data summary may help to suggest a suitable model which will in turn suggest an appropriate inferential procedure. This first phase of the analysis will be described as the 'initial examination of data' or 'initial data analysis' and will be abbreviated to IDA. It has many things in common with **exploratory data analysis** or EDA (Tukey, 1977), but there are also some important differences, as outlined in section 6.9. I regard IDA as an essential part of nearly every analysis and one aim of this book is strongly to encourage its more systematic and thorough use. Although most textbooks cover simple descriptive statistics, students often receive little guidance on other aspects of IDA. This is a great pity. Although IDA is straightforward in theory, it can be difficult in practice and students need to gain experience, particularly in handling messy data and in using IDA as a signpost to inference.

Of course, if you just literally 'look' at a large set of raw data, you will not see very much. IDA provides a reasonably systematic way of digesting and summarizing the data, although its exact form naturally varies widely from problem to problem. Its scope will also depend on the personal preferences of the statistician involved, and so there is no point in attempting a precise definition of IDA. I generally take a broad view of its ingredients and objectives, as becomes evident below. I have found that IDA will often highlight the more important features of a set of data without a 'formal' analysis, and with some problems IDA may turn out to be all that is required (for example, Exercise B.3). Alternatively, IDA may suggest reasonable assumptions for a stochastic model, generate hypotheses to be tested, and generally give guidance on the choice of a more complicated inferential procedure. The important message from all this is that IDA should be carried out **before** attempting formal inference and should help the analyst to resist the temptation to use elaborate, but inappropriate, techniques without first carefully examining the data. Note also that the simple techniques of IDA are particularly relevant to analysts in the Third World who may not have ready access to a computer.

The first part of IDA consists of assessing the structure and quality of the data and processing them into a suitable form for analysis. These topics are covered in sections 6.2–6.4 and may collectively be referred to as **data scrutiny**.

6.2 Data structure

The analysis will depend crucially not only on the number of observations but also on the number and type of variables. Elementary textbooks naturally concentrate on the simple, instructive, but atypical case of a small number of observations on just one or two variables.

As regards the number of observations, any model fitting is likely to be unreliable if the sample size is very small (e.g. less than about 10 for quantitative measurements or less than about 30 observations for categorical variables). It is normally best to treat such a small sample as an exploratory, pilot sample to get ideas. On the other hand, with hundreds, or even thousands, of observations, the problems of data management become severe (section 6.3). Although model fitting is apparently more precise, it becomes harder to control the quality of large data sets, and one must still ask if the data are representative. As already noted in Chapter 4, it is often better to collect an intermediate size sample of good quality rather than a large, but messy, data set. Effective supervision of data collection is crucial.

The number of variables is also important. An analysis with just one or two variables is much more straightforward than one with a large number of variables. In the latter case, one should ask if all the variables are really necessary (they often are not!), or consider whether the variables can be partitioned into unrelated groups. Alternatively, multivariate techniques may be used to reduce the dimensionality (section 6.6). It is potentially dangerous to allow the number of variables to exceed the number of observations because of non-uniqueness and singularity problems. Put simply, the unwary analyst may try to estimate more parameters than there are observations. Indeed, it is generally wise to study data sets where there are considerably more observations than variables and Appendix A.6.5 gives my crude rule of thumb for multiple regression. In some areas (e.g. economics), a large number of variables may be observed and attention may need to be restricted to a suitable subset in any particular analysis.

The type of variable measured is also important (see also Cox and Snell, 1981, section 2.1). One important type of variable, called a **quantitative variable**, takes numerical values and may be **continuous** or **discrete**. A continuous variable, such as the weight of a manufactured item, can (theoretically at least) take any value on a continuous scale. In practice, the recorded values of a continuous variable will be rounded to a specified number of significant figures but will still be regarded as continuous (rather than discrete) unless the number of possible values is reduced to less than about ten. A discrete variable can only take a value from a set or sequence of distinct values, such as the non-negative integers. An example is the number of road accidents experienced by a particular individual in a particular time period.

In contrast to a quantitative variable, a **categorical** variable records which one of a list of possible categories or attributes is observed for a particular sampling unit. Thus categorical data are quite different in kind and consist of the counts or frequencies in particular categories (e.g. Exercise B.4). A categorical variable is called **nominal** when there is no particular ordering to the possible values (e.g. hair colour in Exercise B.4), but is called **ordinal** when there is a natural ordering (e.g. a person's rank in the army). A **binary** variable has just two possible outcomes (e.g. 'success' or 'failure') and is a special type of categorical variable.

The **numerical coding** of categorical data is a tricky exercise fraught with difficulty (section 6.3). In particular, there can be a blurred borderline between ordinal and quantitative variables. For example, in the social sciences many observations are opinion ratings or some other subjective measure of 'quality'. Opinions could be rated in words as very good, good, average, poor or very poor. This is a categorical variable. If the data are then coded from say 1 (very good) to 5 (very poor), then the variable is converted to a discrete form. However, the 'distance' between 1 and 2 may not be regarded as the same as that between, say, 2 and 3, in which case the variable should not be treated as an ordinary discrete quantitative variable. As another example, the continuous variable 'yearly income' in Exercise B.4 is grouped into four categories, such as 'less than 1000 kroner', and this creates an ordinal categorical variable.

To clarify the above remarks, it may be helpful to distinguish between data measured on different types of measuring scales. These include:

- a **nominal** scale, for unordered categorical variables;
- an **ordinal** scale, where there is ordering but no implication of distance between scale positions;

- an **interval** scale, where there are equal differences between successive integers but where the zero point is arbitrary;
- a **ratio** scale, the highest level of measurement, where one can compare the relative magnitude of scores (the ratios) as well as differences in scores.

A measurement on a continuous variable is generally made on a ratio scale provided the measuring scale starts at zero (as it does for height and for temperature in degrees absolute) but is only measured on an interval scale when the scale does not start at zero (for example, for temperature in degrees Celsius). Thus a table 4 metres long is twice as long as a table 2 metres long (the ratio is meaningful), whereas the ratio of say $+15°C$ to $-10°C$ is meaningless. Note that subjective measures of 'quality' are sometimes made over a continuous range, say from 0 to 1 or from 0 to 10 (perhaps by making a mark with a pencil at an appropriate distance along a line). These data look continuous but it could be argued that the measuring scale is actually ordinal. A discrete variable is generally measured on a ratio scale if positive integers are recorded, but on an interval scale otherwise. A categorical variable is generally measured on an ordinal or nominal scale rather than an interval or ratio scale.

The exact meaning of a **qualitative** variable is unclear in the literature. Some authors use 'qualitative' to be synonymous with 'non-numerical' or 'categorical', while others restrict its use to nominal (unordered categorical) variables. Others describe qualitative data as those which cannot be characterized by numerical quantities, or which cannot be averaged in a meaningful way. To avoid confusion, I will generally avoid the term.

The sort of analysis which is suitable for one type of variable or one type of measuring scale may be completely unsuitable for a different type. Problems arise in practice when there is a mixture of different types of variable. It is sometimes possible or necessary to form two or more separate groups of variables of a similar type. Another 'trick' is to turn a continuous variable into an ordinal variable, or even a binary variable, by grouping values into categories. This inevitably leads to some loss of information but may simplify the analysis considerably and, if the results are clear-cut, is fully justified. Examples of the latter are given in Exercises B.4(b) and E.4.

Example 6.1 Problems with measurement scales

Many mistakes are caused by failing to distinguish between different types of variable and different types of measurement scale, and in particular between count and measured data. The following examples are all too typical.

1. *Which is the strongest type of roof support?* Following concern about the collapse of roof supports in mine tunnels, six different ways of holding up the roofs of tunnels were examined. Measurements of strength were obtained for each roof support system, the six systems were labelled 1 to 6, and then a standard ANOVA was carried out. Unfortunately this analysis treated the

type of roof support as if it were measured on an interval scale. No significant differences were found. In fact the type of roof support is clearly a nominal variable. The numbers used to code the different support systems are arbitrary and any other ordering could have been used. In fact one system was much better than the rest, as was clear in the IDA or from an ANOVA when carried out properly.

2. *When is an integer a count?* A company produces long lengths of a certain material. A quality control scheme was set up which included measuring (at the end of each shift) the length y (in metres) of material produced, together with the length of scrap w (in metres). The control scheme assumed that w was a binomial variable because both y and w were recorded as integers. In fact they were measured variables, albeit rounded to the nearest integer. As a result the calculated control limits were much too wide.

It is also important to ask if the variables arise 'on an equal footing' or if instead there is a mixture of **response** and **explanatory** variables. Some techniques (e.g. regression) are concerned with the latter situation in trying to explain the variation in one variable (the response) in terms of variation in other variables (the explanatory or predictor variables). Other techniques are concerned with the former situation. For example, given the different exam marks for different students, the analysis usually consists of some sort of averaging to produce an overall mark. There are also many other more sophisticated, multivariate techniques for examining the interrelationships between a set of comparable variables, e.g. principal component analysis (section 6.6). Note that exam marks might alternatively be regarded as response variables if other (explanatory) information was available about individual students.

Assessing the structure of the data must also take account of the prior knowledge of the system in regard to such matters as the design of the experiment, the known sources of systematic variation (e.g. any blocking factors or known groupings of the experimental units) and so on. An isolated, unstructured data set is quite different from the sort of data arising from a proper experimental design. For example, the difference between hierarchical (nested) and crossed data discussed in Appendix A.11 is fundamental and one must match the model and the analysis to the given problem.

6.3 Processing the data

Apart from 'small' data sets (e.g. less than 20 observations on one or two variables), the data will probably be analysed using a computer. This section considers the problems involved in getting data onto a computer, which is not a trivial exercise. When a set of data is to be assembled for some purpose, then between 25% and 70% of the total effort may be devoted to collecting, examining, processing and cleaning data in order to create a useful data file suitable for analysis. Creating such files poses many challenges. Errors can be hard to catch (section 6.4) and considerable skill is needed to clean data files. There is some help on these topics in the computing literature, but surprisingly little help in statistics texts. Research on topics such as **data cleaning** should be 'rewarding and exciting' (Chambers, 1993), but the prevailing view among academics is unfortunately that such activities are 'tedious' and 'unlikely to yield publishable results'.

Managing data files requires meticulous attention to detail. For example, suppose you have 1 million observations but the counts of each variable reveal one fewer observation on one variable than all the rest. The temptation then is to say that 'it is only one in a million' and so does not matter. In fact it could matter very much and be a tip-of-the-iceberg sign that something serious is wrong. If one observation for one variable is missing half-way through the data file and all remaining observations on that variable have shifted their position by one place, then all correlations and relationships could be ruined.

Some data are recorded directly onto a computer. Examples include measurements from some medical machinery, hospital patient details recorded directly onto a computer rather than on paper, and the use of optical mark recognition (see below). Computer-assisted data recording is likely to increase in future and 'the quality of automatically recorded data, which may never be scanned by human eye, is an issue in urgent need of research' (Gower, 1993).

Other data will typically be recorded manually in data sheets and then need to be (i) **coded** before being (ii) **typed** into a data file and then (iii) **edited** to remove errors. It is arguably a matter of definition as to when a data file becomes a **database**. Some people might say that any collection of numbers on a particular topic can comprise a database, but it may be safer to restrict the term to data files set up using a proper **database management system** (DBMS). A database consists of a set of **records**, each of which comprises the same number of **fields**. Each field will contain a single piece of information which may be of **numerical** or **character** type. The latter may contain any symbol, including numbers and letters (the **alphanumeric** characters), as well as other symbols such as ' + '. Examples include an address (numbers and letters), a surname (all letters) or a telephone number (all numbers, but an initial zero would be chopped off if it were mistakenly entered into a numerical type field). The **width** of a character field will normally be specified by the user, but the width of numerical fields is normally constrained only by the type of computer. A good DBMS will alert the user if he/she tries to enter information into a field when it is of an inappropriate form (e.g. too wide). If a DBMS is not used, then it is usually preferable to set up a simple ASCII file as this is easily transported between different computers and packages.

Data are increasingly made available in the form of a 'ready-made' database over whose construction the statistician may have had little or no control, especially in large companies where databases may be managed by some sort of computer services department. Clearly it is especially important then to check quality. Sometimes a statistician is given what is described as a sample from a database. Without access to the complete database, it may be impossible to check if the sample really is random and representative.

When processing data, the statistician may need to consider the following points:

1. *Choice of variables.* The first step in constructing a data file is to list the variables to be included together with their units of measurement; screen the data to see if all the variables are present and worth including; choose a sensible order for the variables.

2. *Choice of format.* An appropriate format must be selected for recording each variable in the data file. This may include the maximum number of allowed spaces for each observation, the number of decimal places, and whether the observations are to be recorded in a numerical or character type of field. If a good DBMS is used, the format of numerical variables is handled automatically, but if a simple computer file is being set up, then it is advisable to include at least one more digit per variable than is strictly necessary so that there are gaps between numbers when they are printed out. This makes it easier to spot errors in a listing of the data. Moreover many data-reading procedures expect a blank or comma between observations and catastrophic results can arise when a number fills all of its available digits and so 'runs on' from the previous observation.

   Two examples illustrate the disastrous consequences which can result from choosing an inappropriate format:

   - A national company set up a database of its customers over ten years ago. One of the variables recorded was the income of each customer (in pounds sterling). Five digits were allowed for this variable which was (just about) enough at the time, but inflation over intervening years means that a growing number of customers now have six-digit incomes. It was several years before company analysts realized that the first digit of such incomes was simply being chopped off. Thus an income of £125 000 would be truncated to £25 000! The use of a modern DBMS would have obviated this problem as numerical fields would typically not be constrained by a fixed width.

   - A large computer file contained many variables including the populations of different administrative areas in British cities. Six digits were allowed for this variable, which seemed adequate as most areas contained four or five-digit numbers. Unfortunately one area did contain over 100 000 people and, when the data were read in using free format (which requires a blank or comma between numbers), an 11-digit number resulted when the previous reading was erroneously joined onto it. This ruined the analysis until the analyst realized what had happened. Generally speaking, it is unwise to try to read a database in free format.

3. *Coding.* Where necessary, the data are transferred to coding sheets, but note that it is often possible to record data in a suitably coded form in the first place; questionnaire forms should generally be designed to allow this to happen since it is good practice to reduce copying to a minimum. Note that **computer-assisted coding** (CAC) is increasing rapidly, especially for population census and large sample surveys (e.g. Dekker, 1994; Tam and Green, 1994).

Apart from the choice of format (see above), there are two particular problems with coding, namely the treatment of missing observations and the coding of non-numerical information.

If an observation is *missing* for any reason, it must be carefully coded so as to distinguish it from the observations that are available. The use of non-numerical symbols, such as '*' or 'NA' (meaning 'not available'), is the safest approach but could give problems in reading the data when other observations in that field are numerical. In any case there may still be some information available. For example, when data are collected from sample surveys, the coding should distinguish between 'refused to reply', 'don't know', and 'not applicable'. Numerical values such as (-1) or zero or 999 are sometimes used to represent missing values (rightly or wrongly) and the analyst should appreciate that this can be a dangerous strategy. The coded values might wrongly be analysed as ordinary observations giving non-sensical results (see Example 6.2 in section 6.4.4). Note that missing values can occasionally be filled in from external sources and this should obviously be done where possible. Alternatively, it may be possible to **impute** (or estimate) the missing values by fitting a model to the data that have been observed (section 6.4.4).

Many data sets include some variables which are naturally non-numerical in form. Examples include categorical variables and the names of individual respondents in a survey. Such data need to be coded with extra care, if at all. For example, names can be recorded in full, or in some abbreviated form (e.g. first initial and first 6 characters of surname), using a character field. If it is desired to preserve anonymity, names can be coded by giving a code number or code name to each respondent and keeping a master list to allow identification of individuals should that be desired.

Nominal categorical variables, such as 'religion', are particularly tricky to handle. Rightly or wrongly, they are often coded numerically, as numerical data are sometimes much easier to handle on a computer. However, any such coding is entirely arbitrary. For example, if hair colour has 'brown' recorded as '1', 'black' recorded as '2' and 'blond' recorded as '3', then an average value would be meaningless, although some derived statistics, such as the proportion of observations in each category, will be meaningful. It is usually safer to use an alphanumeric code such as 'Br' for brown, 'Bl' for black and 'Bd' for blond.

Ordinal categorical variables have categories with a natural ordering so that numerical coding is not entirely arbitrary. As a result, it is often convenient to code ordinal data numerically. For example, opinion ratings on a five-point scale from 'agree strongly' to 'disagree strongly' could be coded from, say, 1 to 5 (or in reverse direction from 5 to 1, as the direction is arbitrary). This coding puts equal-spaced intervals between successive ratings. If the data are then treated as quantitative (as they often are) by, for example, taking averages, then there is an implicit, and perhaps unjustified, assumption that the scale of measurement is interval. Nevertheless, ordinal data are often treated in this way, especially to get simple descriptive statistics, and it is reasonable provided that the measurement scale is 'nearly interval' as it often is. Certainly this is often much easier than trying to handle non-numerical data.

4. *Typing.* Unless recorded directly into a computer, the coded data have to be typed and stored on tape or disk, either as a simple computer file or as a database using a DBMS. The typist should be encouraged to call attention to any obvious mistakes or omissions on the coding sheets. If possible, the data should be repunched to **verify** the data. Any differences between the two typed versions of the data may then be investigated.

   Note that the use of non-keyboard data-entry methods, such as **optical mark recognition** (OMR), is increasing (e.g. Dekker, 1994; Tam and Green, 1994). The latter scans and reads self-coded responses without human intervention (except for rejected data sheets which the machine cannot read).

5. *Editing.* The data are checked for errors. Comments on data editing will be made in the next section.

6.4 Data quality

The quality of the data is of paramount importance and needs to be assessed carefully, particularly if a statistician was not consulted before they were collected. Are there any suspicious-looking values? Are there any missing observations, and if so why are they missing and what can be done about them? Have too many or too few significant digits been recorded? All sorts of problems can, and do, arise! If, for example, the data have been recorded by hand in illegible writing, with some missing values, then you are in trouble! The main message of this section is that the possible presence of problem data must be investigated and then appropriate action taken as required.

6.4.1 How were the data collected?

The first task is to find out exactly how the data were collected, particularly if the statistician was not involved in planning the investigation. Regrettably, data are often collected with little or no statistical guidance. For example, a poor questionnaire may have been used in a sample survey, or an experimental design may not have been randomized. It is also advisable to find out how easy it is actually to record the data (try counting the number of seedlings in a box!) and to find out exactly what operations, if any, have been performed on the original recorded observations. Problems may also arise when data from several sources are merged. For example, the same variable may have been measured with different precision in different places. More generally, using data from several different sources can, and does, lead to inconsistencies which need to be resolved.

Fortunately, IDA is helpful in assessing what can be salvaged from a set of messy data (e.g. Exercises D.1 and D.2) as well as for actually revealing data-collection inadequacies (e.g. sections 6.4.3 and 6.4.5 and Exercises A.1(d) and D.2). At the other extreme, a proper statistical design may have been used, yielding a highly structured data set. Here the form of analysis may be largely determined a priori, and then IDA may be confined to some simple descriptive statistics and a few quality checks.

6.4.2 Errors and outliers

A result which looks interesting is probably wrong!

The three main types of problem data are errors, outliers and missing observations. This subsection considers the distinction between errors and outliers more closely, while the next subsection deals with ways of detecting and correcting them. Missing observations are considered in section 6.4.4.

An **error** is an observation which is incorrect, perhaps because it was recorded wrongly in the first place or because it has been copied or typed incorrectly at some stage. An **outlier** is a 'wild' or extreme observation which does not appear to be consistent with the rest of the data. Outliers arise for a variety of reasons and can create severe problems. A thorough review of the different types of outlier, and methods for detecting and dealing with them, is provided by Barnett and Lewis (1985).

Errors and outliers are often confused. An error may or may not be an outlier, while an outlier may or may not be an error. Think about this! For example, if a company's sales are recorded as half the usual figure, this may be because the value has been written down wrongly – an error and an outlier – or because there has been a collapse of demand or a labour dispute – giving a true value which is an outlier. These two situations are quite different, but unfortunately it is not always easy to tell the difference. In contrast, if true sales of 870 items, say, are wrongly recorded as 860 items, this error will not produce an outlier and may never be noticed. Thus an outlier may be caused by an error, but it is important to consider the alternative possibility that the observation is a genuine extreme result from the 'tail' of the distribution. This usually happens when the distribution is skewed and the outlier comes from the long 'tail'.

Types of error

There are several common types of error, which are illustrated in Exercises A.1, D.1 and D.2, including the following:

1. A **recording** error arises, for example, when an instrument is misread.
2. A **typing** error arises when an observation is typed incorrectly.
3. A **transcription** error arises when an observation is copied incorrectly, and so it is advisable to keep the amount of copying to a minimum.
4. An **inversion** arises when two successive digits are interchanged at some stage of the data processing, and this is something to be on the look-out for. If, for example, the observation 123.45 appears as 123.54, then the error is trivial, does not produce an outlier, and will probably never be noticed. However, if 123.45 is inverted to 213.45, then a gross outlier may result.
5. A **repetition** arises when a complete number is repeated in two successive rows or columns of a table, thereby resulting in another observation being omitted. More generally, it is disturbingly easy to get numbers into the wrong column of a large table even with the careful use of a spreadsheet package.
6. A **deliberate** error arises when the results are recorded using deliberate falsification, as for example when a person lies about his/her political beliefs.

Some errors are unclassifiable and emphasize the need for eternal vigilance. For example, I recall a set of clinical trial data where several patients were recorded as 'successfully completing treatment' after just one or two days. This seemed medically unlikely, and further enquiries showed that these patients had actually died! As another example, non-metric measurements can often be rounded in a most unfortunate way. An example is recording 5 feet 11 inches as 5.11 feet! Further instructive examples illustrating the choice of an inappropriate computer format and poor coding of missing observations and non-numerical information are given in sections 6.3 and 6.4.4.

6.4.3 Dealing with errors and outliers

The search for errors and outliers is an important part of IDA. The terms **data editing** and **data cleaning** are used to denote procedures for detecting and correcting errors. Generally this is an iterative, ongoing process.

Some checks can be made 'by hand', but a computer can readily be programmed to make other routine checks and this should be done. The main checks are for **credibility**, **consistency** and **completeness**. Checks for missing observations are easy to carry out and methods for dealing with them are described in section 6.4.4. Credibility checks include carrying out a **range test** on each variable. Here a credible range of possible values is prespecified for each variable and every observation is checked to ensure that it lies within the required range. For example, one would normally expect the heights of adult individuals to lie between, say, 1 and 3 metres. These checks will pick up gross outliers as well as impossible values (e.g. height cannot be negative!). Bivariate and multivariate checks are also possible. A set of checks, called 'if-then' checks, can be made to assess credibility and consistency between variables. For example, it is easy to check if age and date of birth are consistent for each individual. While it is generally better to keep the number of variables to a minimum, it is sometimes a good idea to include some **redundant** variables in order to be able to spot errors more easily and also help to fill in missing observations.

Another simple, but useful, check is to get a printout of the data and examine it by eye. Although it may be impractical to check every digit visually, the human eye is very efficient at picking out suspect values in a data array provided they are printed in strict column formation in a suitably rounded form. Suspect values can be encircled as in Table D.1 of Exercise D.1 (see also Chatfield and Collins, 1980, section 3.1). It is certainly worth looking closely at a partial listing of the data (e.g. the first 20 or 30 records). This will give some understanding of the units of measurement, the location and spread of the different variables, the relationships (if any) between variables, the extent of missing data and of non-numerical data, and so on. It can also be helpful at a very early stage to compute the average, maximum and minimum for each variable, though this will of course be done (again) at the later 'descriptive statistics' stage of IDA, when a variety of plots will also be carried out. These plots may enable further suspect values to be detected by eye. In particular, it is important to examine the frequency distribution of each variable, perhaps by means of histograms or stem-and-leaf plots. As well as disclosing possible outliers, the frequency distributions may reveal unexpected gaps between values which could indicate recording problems. Scatter plots of pairs of variables are also helpful in disclosing outliers.

There are many other procedures (e.g. Barnett and Lewis, 1985) for detecting outliers, including significance tests and more sophisticated graphical procedures. Some of these involve looking for large residuals after a model has been fitted and so do not form part of IDA (see section 5.3.3). This is particularly true for multivariate data where outliers may be difficult to spot during the IDA which typically looks only at one and two variables at a time.

When a suspect value has been detected, the analyst must decide what to do about it. It may be possible to go back to the original data records and use them to make any necessary corrections. Inversions, repetitions, values in the wrong column and other transcription errors can be corrected in this way. In other cases, such as occasional computer malfunctions, correction may not be possible and an observation which is known to be an error may have to be treated as a missing observation.

Extreme observations which, while large, could still be correct, are more difficult to handle. The tests for deciding whether an outlier is significant provide little information as to whether an observation is actually an error. Rather external subject-matter considerations become paramount. It is essential to get advice from people in the field as to which suspect values are obviously silly or impossible, and which, while physically possible, are extremely unlikely and should be viewed with caution. Sometimes further data may resolve the problem. It is sometimes sensible to remove an outlier, or treat it as a missing observation, but this outright rejection of an observation is rather drastic, particularly if there is evidence of a long tail in the distribution. Sometimes the outliers are the most interesting observations. I once found that the clinician recording data in a particular clinical trial was discarding all observations outside the range (mean ± two standard deviations) without further considera- tion. This is silly at the best of times but was doubly so here when the data were skewed.

An alternative approach is to use **robust** methods of estimation which automatically downweight extreme observations (section 7.1). For example, one possibility for univariate data is to use **Winsorization**, in which an extreme observation is adjusted towards the overall mean, perhaps to the second most extreme value (either large or small as appropriate). However, many analysts prefer a diagnostic approach which highlights unusual observations for further study (computers typically 'star' them with an asterisk).

My recommended procedure for dealing with outliers, when there is no evidence that they are errors, is to repeat the analysis with and without suspect values (see Exercise B.1, for example). If the conclusions are similar, then the suspect values do not matter. However, if the conclusions differ substantially, then the values do matter and additional effort should be expended to check them. If the matter still cannot be resolved, then it may be necessary to present two lots of results or at least point out that it may be unwise to make judgments from a set of data where the results depend crucially on just one or two observations (called **influential** observations).

Whatever amendments are required to be made to the data, there needs to be a clear, and preferably simple, sequence of steps to make the required changes in the data file.

6.4.4 Missing observations

Missing observations arise for a variety of reasons. A respondent may forget to answer all the questions, an animal may be killed accidentally before a treatment has had time to take effect, a scientist may forget to record all the necessary variables, or a patient may drop out of a clinical trial.

It is important to find out why an observation is missing. This is best done by asking 'people in the field'. In particular, there is a world of difference between observations lost through random events, and situations where missing observations due to damage or loss are more likely to arise under certain conditions. Then the probability that an observation, y, is missing may depend on the value of y and/or on the values of explanatory variables. Only if the probability depends on neither are the observations said to be **missing completely at random** MCAR).

For multivariate data, it is sometimes possible to infer missing values from other variables, particularly if redundant variables are included (e.g. age can be inferred from date of birth).

With univariate data from a proper experimental design, it is usually possible to analyse the data directly as an unbalanced design (e.g. with the GLIM package), but if there are several factors, it may be more helpful to estimate, or **impute**, the missing values by least squares so as to produce a 'fake' fully balanced design. This may help one to understand the data and interpret the results of the ANOVA. In a one-way classification with a single missing value, the latter merely reduces the corresponding group size by one and no imputation is necessary. In a two-way classification (e.g. a randomized block design), a single missing value is replaced by $(tT + bB - S)/[(t - 1)(b - 1)]$, where t is the number of treatments, b the number of blocks, T the sum of observations with the same treatment as the missing observation, B the sum of observations in the same block as the missing observation, and S the sum of all the observations. Then a two-way ANOVA could be carried out in the usual way but with the residual degrees of freedom reduced by one. More generally, there are several such algorithms for replacing missing observations with 'guessed' values so as to allow a 'standard' analysis. However, the textbook example of a single missing univariate observation is quite different from the common problem of having many missing observations in multivariate data, and statisticians have often used a variety of *ad hoc* procedures, such as discarding any multivariate observation which is incomplete. The problem with such *ad hoc* procedures is that bias may result unless the observations really are MCAR (which is unusual). Little and Rubin (1987) have described a more objective approach based on the likelihood function derived from a model for the missing data. This may utilize an algorithm called the E-M algorithm (see Appendix A.4).

Example 6.2 The dangers of coding missing observations

These examples reiterate the remarks from section 6.3 on the dangers of coding missing values with special numerical values such as 999 or zero.

1. *Missing ages.* I once analysed what I thought was a complete set of data concerning the driving records of a sample of female drivers. The results I obtained looked most peculiar until I realized that some ladies had refused to give their age and that all unknown values had been coded as '99'!

2. *Public holidays.* A time series of daily stock prices was given to me for analysis. An initial inspection revealed that values for public holidays had been coded as '999'. These values had to be replaced with an appropriate fitted value before the data could be analysed using standard time-series methods. They could not just be omitted as this would have made it difficult to measure any weekly cyclical variation.

3. *A near disaster.* I was called in as consultant to analyse a large data set comprising 20 000 observations on 20 variables which had already been set up before I arrived. During the analysis we looked at the correlation matrix of the variables and noticed, to our horror, an off-diagonal value of 1.0000000! We tried to think of possible reasons for this. Could one variable have been erroneously duplicated? No, because a quick check of the first few values in the two data columns showed they were not the same. Could one variable be a linear combination of the other? No, because the two suspect variables had different correlations with the other variables. Could the computer program be wrong? No, not for a standard calculation like this.

We next checked the histogram of each variable (which we should have done in the first place, but we were in a hurry) and found to our dismay that there was a very small number of exceptionally extreme values recorded as 999999999 for both variables. Subsequent enquiries revealed that this was the default value for missing observations and the two variables happened to have identical missing values. The small number of duplicated default values was enough to produce the apparently 'perfect' correlation. We hastily excluded the missing observations and continued with the analysis, congratulating ourselves that we had spotted the problem in time. We complained bitterly to the computing department that the selected default value was utterly ridiculous only to be told that they had deliberately chosen it that large so that no one could possibly miss it!! This is an excellent example of the importance of avoiding trouble during statistical investigations, and further details are given in the Prelude of Chatfield (1991).

Of course, rather than just trying to cope with missing observations after they have occurred, it is a good idea to take energetic steps to avoid missing observations, for example by encouraging patients to stay involved in a clinical trial. It is also a good idea to collect covariates for items/subjects which might go missing, so that missing values may be better predicted, and to collect information about the reasons why an observation is missing or a patient has dropped out of a trial.

6.4.5 Precision

As part of the assessment of data quality, it is important to assess the precision of the data. It may appear 'obvious' that if the data are recorded to, say, five significant digits, then that is the given precision. However, this is often not the case. The true recording accuracy may only become evident on arranging the data in order of magnitude, or on looking at the distribution of the final recorded digit.

For a continuous variable, one would normally expect the distribution of the final recorded digit to be roughly uniform (so that all ten digits are approximately equally likely), but this may not be the case. For example, it is common to find that 'too many' observations end in a zero, indicating some rounding. Preece (1981) gives some fascinating examples showing how common-sense detective work on the values of the final digits can reveal a variety of problems, such as difficulties in reading a scale (e.g. Exercise B.3), evidence that different people are measuring to different accuracy, or evidence that the given data have been transformed (e.g. by taking logarithms) or converted from different units (e.g. from inches to centimetres – Exercises A.1 and D.2).

While on the subject of precision, the reader is warned not to be fooled by large numbers of apparently significant digits. In 1956, Japanese statistics showed that 160 180 cameras had been exported to the USA, while the corresponding American statistic was 819 374 imported cameras from Japan. Thus both countries claim six-digit accuracy but cannot even agree on the first digit! The reader may be able to think of reasons for this huge discrepancy, such as imports via a third country, but that is not the point.

6.4.6 Concluding remarks

Data processing and data editing require careful attention to ensure that the quality of the data is as high as possible. However, it is important to realize that some errors may still get through, particularly with large data sets. Thus diagnostic procedures at the later model-building stage should be carried out to prevent a few data errors from substantially distorting the results. With 'dirty' data containing outliers and missing observations, limited but useful inference may still be possible, although it requires a critical outlook, a knowledge of the subject matter and general resourcefulness on the part of the statistician.

6.5 Descriptive statistics

After the data have been processed, the analysis continues with what is usually called descriptive statistics. Summary statistics are calculated and the data are plotted in whatever way seems appropriate. We assume some familiarity with this topic (Appendix A.1 and Exercises A.1 and A.2) and concentrate on comparative issues.

6.5.1 Summary statistics

Summary statistics should be calculated for the whole data set and for important subgroups. They usually include the mean and standard deviation for each variable, the correlation between each pair of variables and proportions for categorical variables. A multi-way table of means (and standard deviations?) for one variable classified by several other variables can also be a revealing exploratory tool.

Measures of location

The **sample mean** is the most widely used statistic, but it is important to be able to recognize when it is inappropriate. For example, it should not be calculated for censored data (Exercise B.5) and can be very misleading for skewed distributions where the **median** (or the **mode** or the **trimmed mean**) is preferred. As a more extreme example, suppose a disease mainly affects young children and old people, giving a U-shaped distribution of affected ages. Then it is silly to calculate the average age of infected people, or indeed any single measure of location.

The average is still widely misunderstood, as indicated by the apocryphal story of the politician who said that it was disgraceful for half the nation's children to be of below average intelligence.

An average by itself can easily be misinterpreted and normally needs to be supplemented by a measure of spread. Thus in comparing journey times, one may prefer a route with a slightly higher mean journey time but smaller variation. Similarly, it is much wiser for a doctor to tell a patient the range of 'normal' blood pressure than to mention a single 'average' value which could cause alarm. This leads us on to measures of spread.

Measures of spread

The **standard deviation** is a widely used measure of variability. Like the mean, it is really designed for roughly symmetric, bell-shaped distributions. Skewed distributions are much harder to describe. I have rarely found measures of skewness and kurtosis to be enlightening. Skewed, bimodal and other 'funny' distributions may be better presented graphically or described in words.

The **range** is sometimes preferred to the standard deviation as a descriptive measure for comparing variability in samples of roughly equal size (especially in quality control), partly because of its simplicity and partly because it is understood much better by non-statisticians. Its lack of robustness is seen as a desirable feature in quality control because outliers are shown up. Unfortunately, direct interpretation of the range is complicated by the tendency to increase with sample size, as described in Appendix A.1. An alternative robust measure of spread with 'nicer' statistical properties, is the **inter-quartile range**.

The **variance** is the square of the standard deviation and is therefore not in the same units of measurement as the data. It should therefore never be used as a descriptive statistic, although it does of course have many other uses. For example, it is the variance which has the property of additivity for independent random variables.

Correlations

Correlations can be useful, but remember that they measure **linear** association, and that people often have difficulty in assessing the magnitude of a correlation and in assessing the implications of a large (or small) value (Exercises C.1 and C.2 and Appendix A.6). Matrices of correlation coefficients, arising from multivariate data, need to be examined carefully, particularly if a technique such as principal component analysis is envisaged (Appendix A.13).

Rounding

The analyst should not give too many significant digits when presenting summary statistics. Ehrenberg's two-variable-digits rule (e.g. Ehrenberg, 1982, Chapter 15; Chatfield, 1983, Appendix D.3) says that data in general, and summary statistics in particular, should be rounded to two variable digits, where a **variable** (or **effective**) digit is defined as one which varies over the full range from 0 to 9 in the kind of data under consideration. Thus given the (fictitious) sample:

$$181.633, \quad 182.796, \quad 189.132, \quad 186.239, \quad 191.151$$

we note that the first digit, 1, is fixed, while the second digit is either 8 or 9. The remaining four digits are variable digits. Summary statistics should therefore be rounded to one decimal place, giving $\bar{x} = 186.2$ in this case.

As a second example, consider the (fictitious) sample:

$$0, \quad 6, \quad 8, \quad 2, \quad 2, \quad 1, \quad 0, \quad 2, \quad 0$$

Here only one digit is recorded and it is a variable digit. Two variable digits are still required in summary statistics and so the mean, for example, should be given with one decimal place, namely $\bar{x} = 2.3$ in this case. Thus an extra digit is given here.

Note that extra working digits may need to be carried during the calculations in order to get summary statistics and other quantities to the required accuracy, so that the two-variable-digits rule does not apply to working calculations.

6.5.2 Tables

It is often useful to present data or summary statistics in a table. The presentation of clear tables requires extra care and the following rules are useful, particularly for two-way tables where numbers are classified by row and by column (as in Exercises B.4 and B.6).

1. Numbers should be rounded to two variable digits (see earlier comments). It is very confusing to present too many significant digits, as usually happens in computer output. Thus computer tables usually need to be revised for presentation purposes.
2. Give row and column averages where meaningful, and perhaps the overall average. Obviously if different columns, say, relate to different variables then only column averages will be meaningful. Sometimes totals or medians, rather than averages, will be appropriate.
3. Consider reordering the rows and the columns so as to make the table clearer. If there is no other natural ordering, then order by size (e.g. order columns by the size of the column averages).
4. Consider transposing the table. It is easier to look down a column than across a row, so that one generally arranges the number of rows to exceed the number of columns.
5. Give attention to the spacing and layout of the table. Keep the columns reasonably close together so that they are easier to compare by eye. With lots of rows, put a gap every five rows or so. Typists often space out a table to fill up a whole page because they think it 'looks better' but a more compact table is usually preferable and needs to be requested before typing commences.
6. Give a clear self-explanatory title. The units of measurement should be stated.
7. Improve the table by trial and error.
8. Give a verbal summary of the major patterns in the table as well as the major exceptions.

Example 6.3 Improving a table

Consider Tables 6.1 and 6.2. Table 6.1 shows part of a table from a published report comparing the accuracy of several different forecasting methods on a

Table 6.1 Average MSE: all data (111)

Method	1	4
NAIVE 1	0.3049E + 08	0.4657E + 09
Holt EXP	0.2576E + 08	0.2193E + 09
BROWN EXP	0.2738E + 08	0.2373E + 09
Regression	0.3345E + 08	0.2294E + 09
WINTERS	0.2158E + 08	0.2128E + 09
Autom. AEP	0.6811E + 08	0.4011E + 09
Bayesian F	0.2641E + 08	0.1328E + 09
Box–Jenkins	0.5293E + 08	0.2491E + 09
Parzen	0.7054E + 08	0.1105E + 09

Table 6.2 Mean square error ($\times 10^7$) averaged over all 111 series (methods ordered by performance at lead time 1)

Method	Forecasting horizon	
	1	4
WINTERS	2.2	21
Holt EXP	2.6	22
Bayesian F	2.6	13
Brown EXP	2.7	24
NAIVE 1	3.0	47
Regression	3.3	23
Box–Jenkins	5.3	25
Autom. AEP	6.8	40
Parzen	7.1	11

large number of time series. The hideous E-format and the lack of ordering makes it impossible to see which method is 'best'. Table 6.2 shows the same results in a suitably rounded form with a clearer title and the methods reordered from 'best' to 'worst' at a forecasting horizon of one period. The results are now much clearer. For example we can see that although Parzen's method is worst at lead time 1, it is actually best at lead time 4. Clear tables like this do not just 'happen', they have to be worked for.

6.5.3 Graphs

'A picture is worth a thousand words.' This old saying emphasizes the importance of graphs at all stages of a statistical analysis.

Most people prefer to look at a graph, rather than examine a table or read a page of writing, and a graphical description is often more easily assimilated than a numerical one. Graphs are ideal for displaying broad qualitative information such as the shape of a distribution (e.g. with a histogram) or the general form of a bivariate relationship (e.g. with a scatter diagram), or data peculiarities. The four graphs in Exercise C.2 illustrate this well. However, tables are often more effective in communicating detailed information, particularly when it is possible that further analysis may be required. In addition, it is worth noting that the features of a graph which make it visually attractive (e.g. colour, design complexity) may actually detract from comprehension. Some people deliberately choose to present fancy graphics which look pretty but are not interpretable. In other words, they use graphics to **hide** information while appearing to do the opposite. The misuse of graphics is particularly hard to combat because most people think they understand graphs.

Descriptive statistics involves plotting a range of simple graphs, usually of one or two variables at a time. It is rather trite to say that one should plot the data in whatever way seems appropriate, but at the exploratory stage one can use the computer to plot a variety of graphs, only some of which turn out to be useful. Different graphs show different aspects of the same data and there is no reason to expect one graph to 'tell all'. The analyst must show resourcefulness and common sense in presenting the data in the best possible way.

The most useful types of graph are histograms, boxplots and scatter diagrams. The (univariate) distribution of each variable should be examined by plotting a **histogram** or a **stem-and-leaf plot** to see what distributional assumptions are reasonable for each variable, and whether any outliers, groupings or other peculiarities are present. An example of a histogram is given in Exercise A.1. Stem-and-leaf plots provide a useful variation on histograms and examples are given in Exercises A.1 and A.2, together with comments on their construction. Stem-and-leaf plots contain more information, but the choice of class interval is a little more restricted. They enable quantiles to be easily calculated and, like histograms, give a quick impression of the shape of the distribution. **Density estimation** (e.g. Silverman, 1986) is a modern non-parametric approach to estimating the underlying probability density function of a distribution. It can be thought of as a way of smoothing the histogram.

A **boxplot** (or box-and-whisker plot) displays a distribution by means of a rectangle, or box, between the upper and lower quartiles, with projections, or whiskers, to the largest and smallest observations. The median is also marked. They are particularly useful for comparing the location and variability in several groups of roughly equal size (e.g. Exercises B.2 and D.2). Non-statisticians often find them a revelation in contrast to a formal analysis of variance.

The **dotplot**, or one-dimensional scatter plot, simply plots each observation as a dot on a univariate scale. It gives fine detail in the tails but does not show up the shape of the distribution very well. It is most effective for small samples up to about size ten, particularly for comparing two or more groups. An example is given in Exercise B.7.

Another general type of graph for examining the shape of a distribution is the **probability plot** (Appendix A.1).

Scatter diagrams are used to plot observations on one variable against observations on a second variable (e.g. Exercise C.2). They help to demonstrate any obvious relationships between the variables (linear, quadratic or what?), to detect any outliers, to detect any clusters of observations, and to assess the form of the random variability (e.g. is the residual variance constant?). Time-series analysis usually starts by plotting the variable of interest against time, and this **time plot** can be regarded as a special type of scatter diagram in which one variable is time.

For multivariate data, it is often useful to plot scatter diagrams for all meaningful pairs of variables. However, with, say, seven variables there are already 21 ($= 7 \times 6/2$) possible graphs to look at. The reader should also realize that it can be misleading to collapse higher-dimensional data onto two dimensions in the form of scatter diagrams or two-way tables when the interrelationships involve more than two variables (see Exercise G.4 for an extreme example). Thus traditional descriptive statistics may only tell part of the story in three or more dimensions. Nevertheless, Exercises D.1 and D.2 demonstrate that a 'simple-minded' approach can often work, even for large multivariate data sets.

As a more sophisticated alternative, there are various procedures for plotting multivariate data which take account of all variables simultaneously. In particular, **Andrews curves** involve plotting a p-variate observation as a function which is a mixture of sine and cosine waves at different frequencies which are scaled according to the values of particular variables. The more controversial **Chernoff faces** represent a p-variate observation as a cartoon face in which each facial feature (e.g. length of nose) corresponds to the value of a particular variable. Some graphs also arise directly from the use of multivariate methods such as principal component analysis (section 6.6).

Although not part of IDA, it is convenient to mention here the wide use of graphs in diagnostic checking (section 5.3.3). Having fitted a model, there are many ways of plotting the fitted values and the residuals. The main idea is to arrange the plots so that under certain assumptions either a straight line will result (to display a systematic component) or a 'random' plot will result (to display a random component).

General remarks

What are the general rules for presenting a 'good' graph?

A graph should communicate information with clarity and precision. It should avoid distorting what the data have to say. It should be capable of making large data sets coherent in a relatively small space. It should encourage the eye to compare different parts of the data. And so on. Some simple rules are as follows:

1. Graphs should have a clear, self-explanatory title. The units of measurement should be stated.
2. All axes should be carefully labelled.
3. The scale on each axis needs to be carefully chosen so as to avoid distorting or suppressing any information. In particular, where a systematic linear trend is expected, the scales should be chosen so that the slope of the line is between about 30° and 45°. The placement of 'tick marks' on the axes also needs careful thought.
4. Use scale breaks for false origins (where the scale on the axis does not start at zero) to avoid being misled. A scale break is indicated by a 'wiggle' on the axis.

5. The mode of presentation needs to be chosen carefully. This includes the size and shape of the plotting symbol (e.g. asterisks or dots) and the method, if any, of connecting points (e.g. straight line, curve, dotted line, etc.).
6. A trial-and-error approach can be very helpful in improving a graph. You are unlikely to get it 'right' first time.

If you still think that plotting a graph is 'easy', then consider the two graphs shown in Fig. 6.1. At first sight they appear to be two different time series, but in fact they are the same time series plotted in two different ways. In the lower graph the vertical scale is stretched out, while the horizontal scale is marked in years rather than months. This makes it much easier to see any seasonal

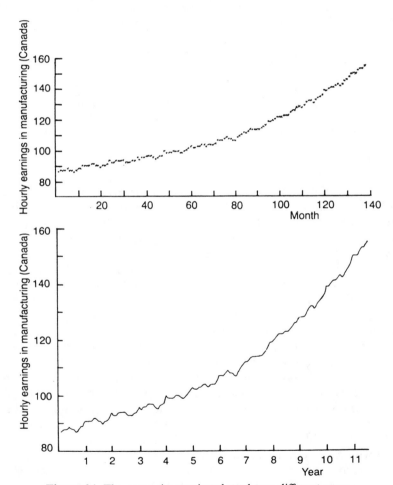

Figure 6.1 The same time series plotted two different ways.

pattern. In addition, the plotted points in the lower graph have been connected by straight lines. Which version, if either, is 'correct'? I think most people will agree that the lower graph looks clearer, but note that neither graph gives the units of measurement for the vertical scale, an omission which needs to be rectified in yet another version. While some graphs are clearly poor, it is not generally helpful to think of a graph as being 'correct' or 'incorrect' since most graphs can be improved with further attention to detail. What is disturbing is that apparently minor changes in presentation can substantially affect the qualitative assessment of the data.

Graphs are widely used in applications, not only in mainstream statistical use, but also in science, social science, and the media. Unfortunately, they are also widely misused, and so statisticians must know how to spot dubious graphics. In particular, it is easy to 'fiddle' graphs by an unscrupulous choice of scales (e.g. Exercise A.4) and the statistician must be on the lookout for this. Tufte (1983) defines what he calls a **lie factor** by

$$\text{lie factor} = \frac{\text{apparent size of effect shown in the graph}}{\text{actual size of effect in the data}}$$

To avoid deception, the lie factor should be kept close to one, but it is unfortunate that many graphs have lie factors near zero or much bigger than one. Tufte (1983, p. 57) gives an example where the lie factor is (deliberately?) chosen to be 14.8!

In his delightful book, Tufte (1983) also suggests avoiding what he calls **chart-junk**. It is all too common to find that graphs have so many labels and comments written all over them that it becomes impossible to see the actual data. Another way of expressing the same message (Tufte, 1983, Chapter 4) is to say that the *ratio of data to ink should be kept high*.

A final piece of general advice is to *use colour sparingly*. Many packages allow colour graphics, and they can look very 'pretty', but most photocopying machines still copy in black and white, so that colour graphs are typically ruined when copied. I have seen this happen all too often.

The above guidelines are not easy to follow unless the analyst has a 'good' software package which allows control of the output. Ideally the analyst should be able to insert appropriate annotation, choose suitable scales, choose an appropriate plotting symbol and so on. Unfortunately, many packages do not give the analyst such control. I have several forecasting packages which produce time plots but which do not allow any alternative specification of scales so that awful graphs often result. Nothing can be done about this other than using correcting fluid and relabelling by hand, or using a better package!

Even with a 'good' package, some hard work is necessary. Figure 6.2 shows the famous airline passenger data given by Box, Jenkins and Reinsel (1994). This time series was first plotted by S-PLUS using the basic default values; Fig. 6.2(a) resulted, which is impossible to read. I then spent some time using

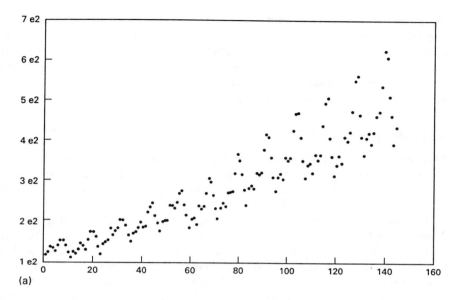

(a)

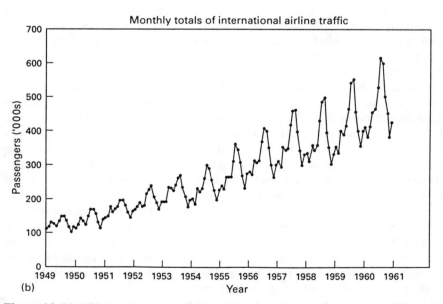

(b)

Figure 6.2 Monthly totals of international airline passengers ('000s): (a) plotted using S-PLUS default options; (b) making use of various optional arguments to get a clearer plot.

the available options to label the axes, add a title, join up the plotted points and so on; Fig. 6.2(b) resulted, which I hope you agree is much clearer. The point is that, even using a 'good' graphics package, the analyst must expend some effort to get a good result. Tufte (1983, p. 92) re-presents a time series drawn in 1786 by William Playfair which is as clear as, if not clearer than, many graphs drawn today. This is a sad reflection on the present computer age!

Interactive graphics

Most graphs, such as histograms and time plots, are **static** and two-dimensional. However we live in a three-dimensional world and most data are multivariate. Thus it is unfortunate that most information displays are stuck in the two-dimensional flatland of paper and computer terminal screens.

In recent years there has been an explosion of interest in **interactive and dynamic graphics** (e.g. Becker, Cleveland and Wilks, 1987) which enable the user to escape from this flatland. The essential feature of interactive graphics is that the data analyst can interact with a two-dimensional graph on a computer screen, for example by manual manipulation of a device such as a 'mouse', so that something happens to the graph. In particular, many tasks can be carried out by a technique called **brushing** in which the analyst uses a mouse to move a rectangle (called a **brush**), across the graph on the computer screen, so as to change some feature of the plot. For example, it may make labels appear and disappear, or eliminate (delete) selected points in order to study the remaining points more easily, or link points on different plots in some obvious way (e.g. by using the same symbol for the same point plotted in different graphs), or highlight selected points by changing them to a darker colour (either in a transient or lasting 'paint mode'). Interactive graphics also allows the analyst to change the scales of different variables, which can be very useful for time plots. A key element of dynamic graphics is that a graph on a computer screen will change according to directions made in a prearranged or interactive way. For example, when looking at three-dimensional clouds of points on a two-dimensional screen, it can be very helpful to rotate the graph about one or more selected axes. Most dynamic graphics are carried out in an interactive way and most interactive graphics packages include dynamic options, so that the two terms are often used interchangeably.

The methods are designed to uncover interesting structures. Subsequently it may be possible to design static displays to show these structures. Thus dynamic graphics is sometimes regarded as a way of moving rapidly through a long sequence of two-dimensional views until an 'interesting' one is found. However, interactive graphics is more than this in that it can uncover features of the data which could not be found using analytic methods.

Sometimes a written paper or book, which has to use two-dimensional graphs, does not convey the excitement of interactive graphics. (This is sometimes called the instamatic-in-Yosemite syndrome. The frustrated photographer can only splutter 'You had to be there to appreciate this!') For some purposes, we are approaching the limits of what a static medium can convey.

Interactive graphics will not only change the way that statisticians explore data, but should also change the way we teach statistics. Analytic methods have dominated statistical teaching but interactive graphical methods offer an alternative approach which can complement and enhance traditional methods. Live visual displays offer insights which are impossible to achieve with traditional methods, and I have seen enough convincing examples to realize that the approach is much more than a mere 'gimmick'. For example, suppose we want to compare two groups of observations. A traditional course would essentially teach students how to compute a confidence interval and carry out a two-sample t-test or some non-parametric alternative. However, a course incorporating interactive graphics would additionally teach students how to compare boxplots and/or dotplots of the two groups. This would include comparing the averages and variability in the two groups, looking for outliers, clusters and distributional shape, and considering the effect of deleting or adjusting observations. It is arguable that students who learn to do this will understand much more about data than those who just calculate a t-value.

Facilities for carrying out interactive graphics are increasing rapidly. Some general software packages, such as S-PLUS, include interactive graphics commands, and there are several more specialized packages such as MACSPIN, Data Desk, REGARD, Diamond Fast and XLISP-STAT. The latter is available free by ftp (file transfer protocol) and will run on both IBM PCs and Apple Macintosh machines.

We give just one example to illustrate the power of interactive methods.

Example 6.4 The use of highlighting for discrimination

In 1936 Sir Ronald Fisher analysed measurements of four variables on 50 specimens of three species of iris, namely *setosa, virginica* and *versicolor*. He used these data to evaluate a technique called **discriminant analysis** which involved finding the best linear function of the four variables to discriminate between the three species. Given a new iris of unknown species we can then measure these four variables (petal width, petal length, sepal width and sepal length), calculate the resulting value of the linear function and apply a rule to this value so as to make an intelligent 'guess' as to which type of iris it is. These data have been reanalysed many times, but many readers may find Unwin's (1992) clear, but simple, analysis based on interactive graphics particularly instructive. Figure 6.3 shows the histograms of the four variables for all 150 observations (50 for each species). In addition, the observations for the *setosa* species have been highlighted. It is now obvious that the widths and lengths of *setosa* petals are much shorter than those of the other two species. A linear discriminant function is not necessary to distinguish between *setosa* and the other two varieties. Rather, a simple rule for deciding whether a new iris of

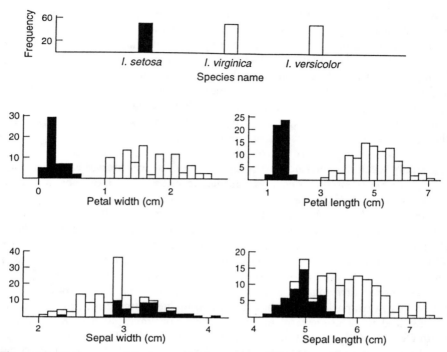

Figure 6.3 Histograms of the four variables measured (in cm) on the iris data with values for *setosa* plants highlighted.

unknown species is *setosa* or not is to say 'an iris of uknown species will be designated as *setosa* if petal length is less than 2.5 cm'. No further analysis is required for this comparison (though it is somewhat more difficult to distinguish between the other two species).

6.5.4 Concluding remarks

Descriptive statistics are useful not only for summarizing a set of data, but also to help check data quality, to start getting ideas for the 'definitive' analysis, and, at the end of the analysis, to help present the conclusions. A number of rules and general principles have been given to ensure that these descriptive statistics are presented clearly. Some of these are apparently self-evident, such as (i) giving graphs and tables a clear, self-explanatory title; (ii) stating the units of measurement; (iii) labelling the axes of graphs; and (iv) rounding summary statistics in an appropriate way. It is a sad commentary on the state of the art that these 'obvious' rules are so often ignored in books, papers, reports and visual display material. To give just one more example, while writing Example

6.4 above, I consulted several books and papers to ascertain the correct units of measurement for the Fisher iris data. I found three books and one paper which did not state the units at all, including one publication which converted the data from centimetres to millimetres without any indication that this had been done. This lack of attention to detail is most unfortunate.

One reason why the 'obvious' rules are disregarded so often, by both statisticians and non-statisticians, is that many tables and graphs are produced by computer packages which find it difficult to imitate the good judgement of a human being as regards such matters as rounding and the choice of scales. For example, the horrible E-format used in Table 6.1 was produced by a computer and needs rounding as in Table 6.2. The upper graph in Fig. 6.1 was also produced by a computer package which divided the horizontal scale automatically into tens. The three graphs in Fig. F.3 in Part Two were produced automatically by a package which did not allow the user to make any change to the (horrible) horizontal scales. Clearly it is important that such computer output be modified in an appropriate way before presenting results and it is regrettable that this is often not done. Of course it would be much more satisfactory to have more control over the output and use a better package in the first place, and this is the ideal we should work towards.

Further reading

There are several good books on graphical methods at a variety of levels including Chambers *et al.* (1983), Tufte (1983) and Cleveland (1985). Some modern computationally-intensive graphical methods are presented in Cleveland (1993), while Becker, Cleveland and Wilks (1987) review dynamic/interactive graphics. Chapman (1986) and Ehrenberg (1982, Chapters 16 and 17) deal with both graphs and tables. Much work needs to be done to give graph construction a proper scientific basis (e.g. Cleveland and McGill, 1987).

6.6 Multivariate data-analytic techniques

Multivariate data consist of observations on several variables. One observation is taken on each variable for each one of a set of individuals (or objects or time points or whatever). Traditional descriptive statistics can be applied to multivariate data by looking at one or two variables at a time, as in section 6.5 above. This is a valuable initial exercise which may prove sufficient as in Exercises D.1 and D.2. In addition, as briefly introduced in section 6.5, there is growing interest in more computationally demanding techniques for presenting multivariate data in two dimensions. These include Andrews curves, Chernoff faces and various forms of interactive graphics.

It is also possible to take a wider view of IDA by allowing the use, where necessary, of a group of more complicated, multivariate techniques which are data-analytic in character. The adjective 'data-analytic' could reasonably be applied to any statistical technique, but I follow modern usage in applying it to techniques which do not depend on a formal probability model except perhaps in a secondary way. Their role is to explore multivariate data, to provide information-rich summaries, to generate hypotheses (rather than test them) and to help generally in the search for structure, both between variables and between individuals. In particular, they can be helpful in reducing dimensionality and in providing two-dimensional plots of the data. The techniques include principal component analysis, multi-dimensional scaling and many forms of cluster analysis (see Appendix A.13 for a brief description and references). They are generally much more sophisticated than earlier data-descriptive techniques and should not be undertaken lightly. However, they are occasionally very fruitful. They sometimes produce results which then become 'obvious' on looking at the raw data when one knows what to look for. They may also reveal features which would not be spotted any other way.

Principal component analysis rotates the p observed variables to p new orthogonal variables, called principal components, which are linear combinations of the original variables and are chosen in turn to explain as much of the variation as possible. It is sometimes possible to confine attention to the first two or three components, which reduces the effective dimensionality of the problem. In particular, a scatter diagram of the first two components is often helpful in detecting clusters of individuals or outliers.

Multi-dimensional scaling aims to produce a 'map', usually in two dimensions, of a set of individuals, given some measure of similarity or dissimilarity between each pair of individuals. This measure could be as varied as Euclidean distance or the number of attributes two individuals have in common. The idea is to look at the map and perhaps spot clusters and/or outliers as in principal component analysis, but using a completely different type of data. Note that psychologists have recently tried to change the data-analytic character of the method by suggesting various probability models which could be applied. However, these models seem generally unrealistic and the resulting analysis does not sustain the informal, exploratory flavour which I think it should have.

Cluster analysis aims to partition a group of individuals into groups or clusters which are in some sense 'close together'. There is a wide variety of possible procedures. In my experience the clusters obtained depend to a large extent on the method used (except where the clusters are really clear-cut) and cluster analysis is rather less fashionable than it was a few years ago. Many users are now aware of the drawbacks and the precautions which need to be taken to avoid irrelevant or misleading results. I often prefer to plot a two-dimensional map of the data, using multi-dimensional scaling or the first two components from principal component analysis, and then examine the graph visually for clusters.

Correspondence analysis is primarily a technique for displaying the rows and columns of a two-way contingency table as points in dual low-dimensional vector spaces. It is the favourite tool of the French *analyse des données* (or data-analysis) school, but is used and understood far less well by English-speaking statisticians. According to your viewpoint, it is either a unifying technique in exploratory multivariate data analysis applicable to many types of data, or a technique which enthusiasts try and force onto all data sets, however unsuitable. The 'truth' probably lies somewhere in-between.

6.7 The informal use of inferential methods

Another way of taking a wider view of IDA is to allow the inclusion of methods which would normally be regarded as part of classical inference but which are here used in an 'informal' way. In my experience, statisticians often use techniques without 'believing' the results, but rather use them in an informal, exploratory way to get further understanding of the data and fresh ideas for future progress.

This type of activity demonstrates the inevitable blurred borderline between IDA and more formal follow-up analyses, and emphasizes the need to regard IDA and inference as complementary. The need to integrate IDA fully into statistics will become even clearer in sections 6.9 and 7.3.

The two main techniques which are used 'informally' are multiple regression and significance tests. The main aim of regression is usually stated as finding an equation to predict the response variable from given values of the explanatory (or predictor) variables. However, I have rarely used regression equations for prediction in a textbook way because of problems such as those caused by correlations between the explanatory variables, doubts about model assumptions and the difficulties in combining results from different data sets. However, I have occasionally found multiple regression useful for exploratory purposes in indicating which explanatory variables, if any, are potentially 'important'.

Significance tests are discussed more fully in section 7.2. Here we simply note that they can be used in an exploratory way even when the required assumptions are known to be dubious or invalid, provided that the analyst uses the results as a rough guide rather than taking them as definitive. It is often possible to assess whether the observed P-value is likely to be an under- or overestimate and hence get informal guidance on the possible existence of an interesting effect, particularly when the result is clear one way or the other.

6.8 Modifying the data

The possibility of modifying the data should be borne in mind throughout the analysis, but particularly at the early stages. In section 6.4 we have already discussed two types of modification to improve data quality, namely:

1. adjusting extreme observations;
2. estimating missing observations.

We therefore concentrate here on two further types of modification, namely:

3. transforming one or more of the variables;
4. forming new variables from a combination of existing variables.

As regards 4, we have already considered, in section 6.6, the possibility of forming a general linear combination of the variables, such as that produced by principal component analysis. Here we have in mind much simpler combinations such as the ratio, sum or difference of two variables. For example, the total expenditure on a particular procedure will almost certainly increase with inflation, and it may well be more informative to consider the derived variable:

$$\begin{array}{ll}\text{(total expenditure} & = \text{(total expenditure)/(some measure} \\ \text{with constant prices)} & \text{of price or the cost of living)}\end{array}$$

Alternatively, it may be better to look at sales in terms of the number of units sold rather than in terms of expenditure. This suggests looking at the derived variable:

$$\text{number sold} = \text{(total expenditure)/(average price per unit)}$$

if the number sold is not available directly. In time-series analysis, a rather different modification is often sensible, namely to take first differences of the given series to see if the change in one variable is related to the change in another variable. Other forms of differencing can also be useful.

As regards transformations, there are many forms in regular use. For example, in scaling exam marks it may be desirable to adjust the mean value and/or the spread by making a transformation of the form:

$$\text{scaled mark} = a + b \quad \text{(raw mark)}$$

where a and b are suitably chosen constants. This is an example of the general linear transformation $y = a + bx$. Common non-linear transformations include the logarithmic transformation given by $y = \log x$ and power transformations such as $y = \sqrt{x}$ and $y = x^2$.

There are various reasons for making a transformation, which may also apply to deriving a new variable:

1. to get a more meaningful variable (the best reason!);
2. to stabilize variance;
3. to achieve normality (or at least symmetry);
4. to create additive effects (i.e. remove interaction effects);
5. to enable a linear model to be fitted.

While it is often helpful to try out different transformations during an IDA, some of the above objectives are really more concerned with model formulation, but it is nevertheless convenient to consider transformations here. One major problem is that the various objectives may conflict and it may prove impossible to achieve them all at once. Another problem is that the general form of a model may not be invariant under non-linear transformations.

One general class of transformations is the Box–Cox family of power transformations given by:

$$y = \begin{cases} (x^{\lambda} - 1)/\lambda & \lambda \neq 0 \\ \log x & \lambda = 0 \end{cases}$$

This is essentially a power transformation, $y = x^{\lambda}$, which is 'fiddled' so that it incorporates the logarithmic transformation as a special case. This follows from the result that $(x^{\lambda} - 1)/\lambda \to \log x$ as $\lambda \to 0$.

The transformation depends on the value of λ, and it is possible to choose λ by trial and error so as to achieve some desired property (e.g. normality) or estimate λ more formally so as to maximize a given criterion (e.g. likelihood). However, in order to get a meaningful variable, it is better to choose λ to be a 'nice' number such as $\lambda = 1$ (no transformation), $\lambda = \frac{1}{2}$(square root) or $\lambda = 0$ (logs). There may be severe interpretational problems if one ends up with, say, $\lambda = 0.59$.

An example where a square root is meaningful is given by Weisberg (1985, p. 149), when examining the relationship between the perimeter and area of 25 Romanesque churches. A 95% confidence interval for λ which maximizes the likelihood for a linear model relating perimeter and (area)$^{\lambda}$ is the interval $0.45 \leq \lambda \leq 0.80$, and then it is natural to choose $\lambda = 0.5$ – the square root – as $\sqrt{}$(area) is in the same units of measurement as perimeter.

Logarithms are often meaningful, particularly with economic data when proportional, rather than absolute, changes are of interest. Another application of the logarithmic transformation is given in Exercise B.9 to transform a severely skewed distribution to normality.

Finally, it is worth stressing that the most meaningful variable is often the given observed variable, in which case a transformation should be avoided if possible. The application of a non-linear transformation, such as the square root, can radically change the character of a set of data and make the resulting inferences misleading, even though the data may perhaps be 'more normal'. In fact the need for transformations has reduced somewhat since computer software now allows more complicated models to be fitted to the raw data. For example, GLIM allows one to fit a general linear model with a gamma 'error' distribution as well as a normal distribution. Transformations should be the exception rather than the rule.

6.9 The importance of IDA

Some readers may be surprised that this section on IDA is so comparatively long. Yet the importance of IDA is one of the main themes of this book, and interest in descriptive data-analytic methods has increased in recent years. This is partly because of dissatisfaction with the overemphasis on more formal mathematical statistics which was prevalent before about 1975, and partly because computational advances have made IDA much easier to do.

We have seen in section 5.1 that statistical analysis has four main phases, namely:

1. descriptive statistics;
2. formulating a sensible model;
3. fitting the model;
4. checking the model.

The two main objectives of IDA are to help in **data description** and to make a start in **model formulation**. Thus IDA is vital in phases 1 and 2 above, and the importance of IDA is clear once it is realized that inference is not just model estimation. Often the problem is not 'How can this model be fitted?' – which is often easy to answer – but rather 'What model, if any, is appropriate here?'. Graphs and summary statistics will help to suggest what **primary** and **secondary** assumptions are reasonable (see section 5.3), as for example in Exercise 5.1 where a scatter plot was used to formulate a sensible regression model. Likewise a time plot is a vital component of time-series model building (Appendix A.14), while in multivariate analysis it is always a good idea to have a preliminary look at the correlation matrix of a set of variables before attempting factor analysis or principal component analysis (Appendix A.13).

Returning to the first objective of IDA, namely data description, it is clear that one should always begin by scrutinizing, summarizing and exploring data. However some aspects are not widely appreciated. In particular, IDA may be all that is required for the following reasons:

1. The objectives are limited to finding descriptive statistics. This usually applies to the analysis of entire populations, as opposed to sample data, and may also apply to the analysis of large samples where the question is not whether differences are 'significant' (they nearly always are in large samples), but whether they are interesting. In addition, IDA is all that is possible when the data quality is too poor to justify inferential methods which perhaps depend on unfulfilled 'random error' assumptions. An IDA may also be sufficient when comparing new results with previously established results. Thus, whereas inference is primarily useful for one-off random samples, IDA can be applied additionally to 'dirty' data and to the analysis of several related data sets.

It should also be noted that the demand on statisticians can be very different in Third World countries. There the emphasis is generally not on extracting the fine detail of data using sensitive inferential techniques, but rather on finding a concise description of general trends and patterns. Then IDA may be perfectly adequate.
2. The results from the IDA indicate that an inferential procedure would be undesirable and/or unnecessary. This applies particularly when significance tests are envisaged but the results of the IDA turn out to be 'clear-cut' or to indicate problems with model assumptions.

The above remarks explain why I have avoided the alternative term 'preliminary data analysis', in that one important message is that an IDA may be sufficient by itself. However, it should also be said that it is not always easy to decide beforehand when a descriptive analysis alone is adequate (Cox and Snell, 1981, p. 24). Sometimes inference adds little or nothing to the IDA, but sometimes it adds a lot. While *ad hoc* descriptive analyses are sometimes very effective, the results lack measures of uncertainty and cannot easily be validated. Thus some statisticians prefer always to carry out a 'formal' model-based analysis to retain 'objectivity'. However, this has dangers and difficulties of its own, notably that it requires the choice of a model, which usually involves subjective judgement. This leads us back to the second main objective of IDA, namely its use in model formulation. Put crudely, an IDA helps you do a 'proper' analysis 'properly'.

6.9.1 Arguments against discussing IDA

Many experienced statisticians have long recognized the importance of IDA. Unfortunately, the literature suggests that IDA is still undervalued, neglected or even regarded with disfavour in some circles. It is therefore worth looking briefly at some arguments which have been put forward against the discussion and use of IDA.

Some people might argue that IDA is all common sense, that scientists do it naturally, and that the subject is too straightforward and well understood to warrant serious discussion. However, I would argue that none of these is true especially when IDA has the wider ingredients and objectives suggested here.

Secondly, IDA is sometimes seen as being *ad hoc* and not based on a sound theoretical foundation. However, I would stress that a lack of theory does not imply that the topic is trivial. Rather, IDA can be more demanding than many classical procedures which have become very easy (perhaps too easy!) to perform with a computer. In fact, much of IDA is not *ad hoc* in that it can be tackled in a reasonably systematic way. However, some aspects of IDA (like some aspects of inference) are *ad hoc*, but this is not necessarily a bad thing. A good statistician must be prepared to make *ad hoc* modifications to standard procedures in order to cope with particular situations. The term *ad hoc* sounds vaguely suspect but is defined as being 'arranged for a special purpose' which is often very sensible.

Finally, IDA may be seen as being dangerously empirical and of downplaying prior knowledge and statistical theory. However, I hope I have said enough to reassure you that while IDA rightly emphasizes the inspection of data, I have no intention of downplaying theory. Rather, I wish to use the data to build on existing theory and fully integrate IDA into statistics.

Of course, analyses based on no model at all do run the risk of giving invalid conclusions, and one must beware of simplistic analyses which overlook important points. However, a simple analysis need not mean a naive analysis, and an IDA should be helpful in deciding when a more complicated analysis is required and of what form. The other side of the coin is that analyses based on the wrong model are likely to be invalid, so that the use of a model does not automatically make things respectable. Nevertheless, it is generally preferable to work within the framework of a probability model, and then IDA can be vital in selecting an appropriate model. On the other hand, there are some occasions when it is fruitful to work without a model, in which case IDA is very important in a quite different way.

In summary, the suggested drawbacks of IDA are far outweighed by the benefits.

6.9.2 *A comparison with EDA*

The initial examination of data is often called **exploratory data analysis** (EDA) after the title of the book by Tukey (1977). The latter describes a variety of graphical and numerical techniques for exploring data. There is no doubt that Tukey's book has provided a major stimulus to data analysis and has been instrumental in adding several new devices to the data analyst's toolkit, notably the stem-and-leaf plot and the boxplot. Unfortunately, Tukey's book can also be criticized for introducing too much new statistical jargon, for suggesting too many new procedures which are rather elaborate for a preliminary analysis, for omitting some standard tools such as the arithmetic mean, for saying very little on the crucial topics of data collection and data processing, and for making little attempt to integrate EDA into mainstream statistics (see also Chatfield, 1985, section 5; 1986). Generally Tukey (1977) shows *how* different tools are used but not *why*, and there is little guidance on the crucial topic of using descriptive statistics in model formulation. For some people **data analysis** is a separate topic to **statistics**, but my view is that it should be an important subset. Thus, to compare IDA with EDA, I suggest that a subset of EDA is a subset of IDA but that the overlap is not as large as might be expected.

There are several alternative books on EDA such as McNeil (1977), Erickson and Nosanchuk (1992), Velleman and Hoaglin (1981), and Hoaglin, Mosteller and Tukey (1983). They have only varying degrees of success in terms of the important task of integrating EDA into mainstream statistics, and I suggest more effort is still needed in this direction. Of course Tukey himself is aware of these wider issues, but they are not evident in his 1977 book. This explains why I have used the alternative title of IDA to emphasize the differences with EDA, both in the details of techniques and in the objectives. The title also emphasizes the importance of integrating descriptive methods into statistics and of using them in model formulation.

6.9.3 Summary

IDA is useful both for scrutinizing and summarizing data and also helps formulate a sensible model. Sometimes IDA may obviate the need for more sophisticated techniques when the results are clear-cut or indicate that the data are of too poor quality to warrant 'proper' inference. The methods are not trivial but require sound subjective judgement. Of course it is not suggested that model-based inference be abandoned, since a good model is often vital, but I do suggest that we need a better balance between IDA and inference both in teaching and in practice.

Further reading

IDA is reviewed by Chatfield (1985), and the lively discussion which ensued is also worth reading.

7

Analysing the data — 3: The 'definitive' analysis

I know **how** to do a t-test, but not **when**!

Although IDA is important, and occasionally sufficient by itself, it should normally be seen as a stepping stone to a proper model-based analysis, which, for want of a better expression, we will call the 'definitive' analysis. This analysis will normally be based on some kind of probability model and will involve appropriate inferential procedures. These may include the estimation of model parameters and the testing of one or more hypotheses.

This chapter briefly introduces some different types of statistical procedure (without going into details – see Appendix A), and then attempts to give some general advice on significance tests and on how to choose the most appropriate procedure for some particular situation. This choice is further complicated by the possibility of using a **robust** or **non-parametric** approach (section 7.1) and by the various possible philosophical approaches to inference. Some general comments on the latter choice are given in section 7.4.

Inference is usually based on a **probability model**. It is important to recognize both the advantages and the limitations of working with a model. Formulating a 'good' model is a key stage of any analysis (section 5.3), but remember that models are always tentative and approximate. It may sometimes be helpful to think of a model in the form (section 5.3.3):

$$DATA = FIT + RESIDUAL$$

An alternative symbolic equation, familiar to engineers, is:

$$DATA = SIGNAL + NOISE$$

Here the signal is the **systematic** component, on which interest is usually centred, while the noise is the **random** component. Note that the systematic component may be a fixed or deterministic function, but there is increasing interest in components which may evolve through time. Also note that the random component may include both **measurement error** and **natural random variability** between the sampling units on which measurements are taken. Statistical analysis can be described as extracting information about the signal in the presence of noise.

7.1 Different types of statistical procedure

There is a bewildering range of statistical procedures, and brief details of some of them are given in Appendix A. Each procedure is applicable for use with a particular type of data for a particular situation. With computer packages readily available, it is no longer so important to remember exact details of the different methods, but rather to understand in general terms what they do and why. This will help the analyst choose an appropriate procedure. For example, the one-sample t-test is designed to assess whether a given sample mean is a 'long way' from a suggested population mean, and I prefer to explain (and remember) the formula for calculating a t-value as being

(difference between sample mean and population mean)/SE (difference)

(where SE means standard error) rather than the mathematical formula

$$t = (\bar{x} - \mu)\sqrt{n}/s$$

in the usual notation; see Equation A.5.2.

There follow brief notes on some different classes of procedure:

1. *Single-sample location problems.* Given a single sample of (univariate) observations, what is the underlying population mean? The estimation problem here would be to find a point estimate or (for preference) a confidence interval for the unknown mean. A significance testing problem would be to assess whether the population mean could equal a particular prespecified value. The estimation and testing of a proportion can be regarded as a special case of the above by thinking of a proportion as the mean of a variable taking the values one or zero according to whether a characteristic is present or not.
2. *Two-sample location problems.* Given samples from two groups or populations, what is the difference between the two underlying population means? Are the two sample means significantly different from one another?
3. *Other significance tests.* There are now literally hundreds of different tests described in the literature and this is potentially very confusing. In view of serious concern among statisticians about their overuse and misuse, section 7.2 gives some general comments and advice on their practical pitfalls. Estimation should be regarded as being more important than hypothesis testing in many situations.
4. *Regression problems.* Given observations on a response variable, y, and several explanatory variables, $x_1, x_2, ..., x_k$, find a regression curve to predict y from the x's. Here the analyst should avoid the temptation to include too

many x's which can give a spuriously good fit. The primary and secondary assumptions made in the model should be checked, both before the analysis (look at scatter plots to see what looks reasonable), and in the later diagnostic-checking phase. The fitted equation is much less reliable when the predictor variables are uncontrolled (as is generally the case with economic data) than when they can be controlled in a proper experimental design. Further aspects of, and problems with, regression are discussed in Appendix A.6.

5. *Factor analysis and principal component analysis.* Given observations on several variables which arise 'on an equal footing', find new derived variables which may be more meaningful in some sense. While sometimes useful, factor analysis in particular is sometimes overused or misused (Appendix A.13).

6. *Analysis of variance* (ANOVA). Given observations from an experimental design, ANOVA partitions the total variation in the response variable (the total corrected sum of squares) into components due to different effects. These effects may include the treatment effects of particular interest (e.g. is a new drug better than existing drugs?), the effects of nuisance factors and of blocking factors, and of any residual (random?) variation. The ANOVA will provide an estimate of the residual variance (the residual mean square) which will in turn allow the testing of hypotheses about the systematic effects and the calculation of confidence intervals for differences in treatment effects. Thus ANOVA leads to an analysis of means, as well as being an analysis of variability.

7. *Other procedures.* There are many other types of statistical procedure such as the analysis of time-series data, of survival data and of categorical data.

Rather than go into more detail on particular techniques, it is more important to realize that many problems can be solved by 'standard' textbook solutions while others need some modification to be made, and that a 'good' method of analysis should:

1. use all relevant data, but recognize their strengths and limitations;
2. consider the possibility of transforming, or otherwise modifying, the given observed variables;
3. try to assess a suitable model structure;
4. investigate the model assumptions implicit in the method of analysis;
5. consider whether the fitted model is unduly sensitive to one or more 'influential' observations.

It is also important to decide if you are going to use a classical parametric model-fitting approach, or use a non-parametric or a robust approach.

A **non-parametric** (or **distribution-free**) approach (Appendix A.5.5) makes as few assumptions about the distribution of the data as possible. It is widely used for analysing social science data which are often not normally distributed, but rather may be severely skewed.

Robust methods (Appendix A.4) may involve fitting a parametric model but employ procedures which do not depend critically on the assumptions implicit in the model. In particular, outlying observations are usually automatically downweighted. Robust methods may therefore be seen as lying somewhere between classical and non-parametric methods.

Now we know that a model is only an approximation to reality. In particular, the fit can be spoilt by (a) occasional gross errors, (b) departures from the (secondary) distributional assumptions, for example because the data are not normal or are not independent, (c) departures from the primary assumptions. 'Traditional' statisticians usually get around (a) with diagnostic checks, where unusual observations are isolated or 'flagged' for further study. This can be regarded as a step towards robustness. Many statisticians prefer this approach except for the mechanical treatment of large data sets where human considera-tion of individual data values may not be feasible. Then it may well be wise to include some automatic 'robustification' as this gets around problem (a) above as well as (b) to some extent. Some statisticians prefer a robust approach to most problems on the grounds that little is lost when no outliers are present, but much is gained if there are. In any case it can be argued that the identification of outliers is safer when looking at the residuals from a robust fit. Outliers can have a disproportionate effect on a classical fit and may spoil the analysis completely. Thus robust procedures are likely to be used more in future when they can be routinely implemented by computer packages.

Non-parametric methods get around problem (b) above and perhaps (a) to some extent. Their attractions are that (by definition) they are valid under minimal assumptions and generally have satisfactory efficiency and robustness properties. Some of the methods are tedious computationally (try calculating ranks by hand) although this is not a problem if a computer is available. However, non-parametric results are not always so readily interpretable as those resulting from parametric model fitting. They should thus be reserved for special types of data, notably ordinal data or data from a severely skewed or otherwise non-normal distribution.

An IDA may help to indicate which general approach to adopt. However, if still unsure, it may be worth trying more than one method (section 7.3).

7.2 Significance tests

A **significance test** may be described as a procedure for examining whether a set of data is consistent with some previously specified hypothesis, called a **null hypothesis**. The general terminology is given in Appendix A.5 together with brief details of some specific tests. Significance tests do have a valuable role, but this role is more limited than many scientists realize, and it is unfortunate that tests are widely overused and misused in many scientific areas, particularly in medicine, biology and psychology. This section gives some advice on practical pitfalls and discusses where their use is inappropriate.

Two statements which I like to stress are:

1. A significant effect is not necessarily the same thing as an interesting effect.
2. A non-significant effect is not necessarily the same thing as no difference.

With regard to statement 1, results calculated from large samples are nearly always 'significant' even when the effects are quite small in magnitude. Thus there is no point in testing an effect which is not substantive enough to be of interest. Thus before doing a test, always ask if the effect is large enough to be of any practical interest. If not, there is little or no point in doing a test. Tests were originally devised for use with small, but expensive, scientific samples. As to statement 2, a large effect, of real practical interest, may still produce a non-significant result simply because the sample is too small. Here an understanding of type II errors and power is useful. It can be disturbing to find how low the power of a particular test is for the size of difference which is of interest.

One basic problem is that scientists often misinterpret a P-value to mean the probability that a null hypothesis, H_0, is true. This is quite wrong. A 'significant' result does not provide 'proof' that a null hypothesis is incorrect, but only evidence of a rather formal type for assessing the hypothesis. This evidence may, or may not, be helpful. Likewise a non- significant P-value does not 'prove' the null hypothesis to be true (Sir Ronald Fisher once wrote that errors of the second kind are committed only by those who misunderstand the nature and application of tests of significance. However, this is a rather extreme view given that 'accepting' a hypothesis need not imply that it has been 'proved'.) Many statisticians, who normally adopt a frequentist approach, are unhappy with the whole machinery of significance testing and the potentially misleading nature of P-values. In contrast, the Bayesian approach (section 7.4) *can* be used to assess the probability that H_0 is true using Bayes' theorem provided that it is possible to assess the prior probability that H_0 is true. The latter assessment is not easy. Strictly speaking, it should be made before seeing the data (though sometimes it is not!), and will depend on the perceived trustworthiness of 'theory' and/or on previous data. There is a big difference between having no previous data and having lots of prior empirical knowledge gathered under a range of conditions. In the latter case, the analyst should probably be looking for interesting discrepancies from H_0 rather than trying to 'reject' H_0.

Another problem with interpreting P-values is the rigid enforcement by some analysts of 5% and 1% probability levels, when they should often be more interested in the qualitative consequences of a model. Thus P-values of 0.04 and 0.06 are essentially saying the same thing even though one is (just) significant at the 5% level and the other is not. On the other hand, P-values of say 0.4 and 0.01 are qualitatively and practically different.

The overemphasis on significance testing, and the rigid interpretation of P-values, has many unfortunate consequences. In particular, it can be difficult to get non-significant results published, leading to what is called **publication bias** (e.g. Begg and Berlin, 1988). This is particularly disturbing when it stops non-significant results being published from confirmatory studies set up to investigate an earlier significant result, especially in medical applications where it can be just as important to know that a new drug does not have a significant effect as to find that it does. The search for 'significance' can also lead an investigator to perform tests on features picked out as unusual *after* performing a descriptive analysis. This includes performing tests at selected points in time or between selected subgroups where the results happen to look 'different'. This is dangerous since the analyst is effectively performing lots of tests without allowing for the effects of multiple testing. Some investigators carry out lots of tests anyway but only report the 'significant' P-values. This is also misleading.

Of course, a single experiment may be just a small part of a continuing study, and yet the literature gives overwhelming emphasis to the idea of looking for a 'significant' effect in a single sample. It is often preferable to see if 'interesting' results are **repeatable**, or generalize to different conditions, by searching for what Nelder (1986) calls **significant sameness** (see also sections 5.2, 7.4).

A related general point is that the **estimation** of effects is often more important than significance testing. Null hypotheses are often rather silly and obviously untrue. For example, it is highly unlikely that two treatments will have *exactly* the same effect, and I have been asked to test such silly hypotheses as 'Water has no effect on plant growth' and 'Summer and winter have the same effect on human health'. Quoting a P-value is generally not the right place to stop. Especially when a result is significant, the analyst should go on to evaluate the implications of the results and assess how large the effect is. This can often be done by means of a confidence interval.

I regard IDA as an important prelude to significance testing, both in generating sensible hypotheses in a first-time study, in checking or suggesting what secondary assumptions are reasonable, and more importantly in indicating that a test is unnecessary, inappropriate or otherwise undesirable for the following reasons:

1. The IDA indicates that the results are clearly significant or clearly not significant. For example, two large samples which do not overlap at all are 'clearly' significant. (In fact by a permutation argument, non-overlapping samples of sizes four or more are significantly different.) Exercise B.3 provides another example of 'clear' significance. In contrast, I was once asked to test the difference between two sample means which happened to be identical! It is impossible to think of a clearer non-significant result. The inexperienced analyst may nevertheless have difficulty in deciding when a test result is 'obvious' and may find that intuition is strengthened by carrying out tests which are unnecessary for the more experienced user.

2. The IDA indicates that the observed effects are not large enough to be 'interesting', whether or not they are 'significant'. This may appear obvious but I have frequently been asked to test results which are of no possible consequence.

3. The IDA indicates that the data are unsuitable for formal testing because of data contamination, a lack of randomization, gross departures from necessary secondary assumptions, inadequate sample sizes, and so on. For example, one or more gross outliers can 'ruin' the analysis (see, for example, Exercise B.1), while a lack of randomization can lead to bias. Other potential problems include a skewed or otherwise non-normal 'error' distribution (Exercise B.7) and a non-constant 'error' variance. It may be possible to overcome some problems by 'cleaning' the data to remove outliers, by transforming one or more variables or by using a non-parametric test.

A different type of problem arises through **interim** and **multiple** tests. If data are collected sequentially, it is tempting to carry out interim tests on part of the data, but this can be dangerous if overdone as there is increasing risk of rejecting a null hypothesis even when it is true. Similar remarks apply when a number of different tests are performed on the same data set. Suppose we perform k significance tests each at the $100\alpha\%$ level of significance. If all the null hypotheses are actually true, the probability that at least one will be rejected is larger than α, and as a crude approximation is equal to $k\alpha$, even when the test statistics show moderate correlation (up to about 0.5). The Bonferroni correction suggests that for an overall level of significance equal to α, the level of significance for each individual test should be set to α/k.

Finally, we consider the general question as to whether it is sound to generate and test hypotheses on the same set of data. In principle, a significance test should be used to assess a null hypothesis which is specified before looking at the data, perhaps by using background theory or previous sets of data. However, tests are often not performed in this 'proper' confirmatory way, and Cox and Snell (1981, section 3.7) discuss the extent to which the method of analysis should be fixed beforehand or allowed to be modified in the light of the data. While it is desirable to have some idea how to analyse data before you collect them, it is unrealistic to suppose that the analysis can always be completely decided beforehand, and it would be stupid to ignore unanticipated features noticed during the IDA. On the other hand, there is no doubt that if one picks out the most unusual feature of a set of data and then tests it on the same data, then the significance level needs adjustment as one has effectively carried out multiple testing. Thus it is desirable to confirm an effect on two or more data sets (see also sections 5.2 and 5.3.3), not only to get a valid test but also to get results which generalize to different conditions. However, when data are difficult or expensive to obtain, then some assessment of significance in the original data set can still be valuable. As this sort of thing is done all the time (rightly or wrongly!) more guidance is badly needed.

Further reading

The role of the significance test is further discussed by many authors including Cox (1977), Cox and Snell (1981, section 4.7) and Morrison and Henkel (1970), the latter containing articles by different authors on a variety of problems. Carver (1978) goes as far as recommending the abandonment of all significance testing in favour of checking that results can be reproduced. Fascinating accounts of the differences between the Fisherian approach to significance testing and the Neyman–Pearson approach to hypothesis testing are given by Lehmann (1993) and Inman (1994).

7.3 Choosing an appropriate procedure

Choosing the appropriate form of analysis can be difficult, especially for the novice statistician. Of course 'experience is the real teacher' but one aim of this book is to hasten this learning process.

We assume that an IDA has already been carried out and that the conclusions are not yet 'obvious'. The first question is whether the form of the definitive analysis has been specified beforehand. If so, does it still look sensible after the first look at the data? Of course it is prudent to specify an outline of the analysis before collecting the data, but the details often need to be filled in after the IDA. In particular, the data may exhibit unexpected features of obvious importance which cannot be ignored.

Suppose instead that the exact form of the analysis has not been specified beforehand. How then do we proceed? The sort of questions to ask are:

1. What are the objectives? You should at least have some broad idea of what to look for.
2. What is the structure of the data? Are they univariate or multivariate? Are the variables continuous, discrete, categorical or a mixture? Look out especially for rank data, for nominal categorical data and for censored data which need extra care. What are the important results from the IDA?
3. What prior information is available? Have you tackled a similar problem before?
4. If not, do you know someone else who has? Ask for help, either within your organization or perhaps at a neighbouring college or research centre.
5. Can you find a similar type of problem in a book? You need access to an adequate library.
6. Can you reformulate the problem in a way which makes it easier to solve? Can you split the problem into disjoint parts and solve at least some of them?

Some general comments are as follows:

(a) You should be prepared to try more than one type of analysis on the same data set. For example, if you are not sure whether to use a non-parametric or parametric approach, try both. If you get similar results, you will be much more inclined to believe them. If, however, the conclusions differ, then more attention must be paid to the truth of secondary assumptions.

(b) It is a mistake to force an inappropriate method onto a set of data just because you want to use a method you are familiar with.

(c) You must be prepared to look at a problem in a completely different way to the one which is initially 'obvious'. In other words, a good statistician must be prepared to use what de Bono (1967) has called 'lateral thinking'. For example, you may want to answer a different question to the one that is posed, or construct different variables to the ones that have been observed.

(d) You must be prepared to make *ad hoc* modifications to a standard analysis in order to cope with the non-standard features of a particular problem. For example, a time-series/forecasting analysis may be helped by discarding the early part of the data if their properties are different to those of more recent data. The presence of outliers and missing observations will also require careful non-standard handling.

(e) You cannot know everything. Thus you are certain to come across situations where you have little idea how to proceed, even after an IDA. For example, you may not have come across censored survival data as in Exercise B.5. There is nothing shameful about this! However, you must know where to look things up (books, journals, reference systems – see Chapter 9) and you must not be afraid to ask other statisticians for help.

(f) 'The analysis' should not be equated with 'fitting a model' (or with estimating the parameters of a given model). Rather 'the analysis' should be seen as a model-building exercise wherein inference has three main strands of model formulation, estimation and model checking. In a first-time study, the model builder's main problem is often not how to fit an assumed model – to which there is often a nice straightforward reply – but rather what sort of model to formulate in the first place. The general remarks in section 5.3 may be worth rereading at this point. The importance of checking the fitted model also needs re-emphasizing. In particular, discrepancies may arise which question the original choice of model or analysis.

(g) Many models (and by implication the appropriate analysis) are formulated to some extent on the basis of an IDA. Following on from Chapter 6, we can now clarify this role for IDA. In many problems a general class of models is entertained beforehand using prior theoretical and empirical knowledge. Nevertheless, the IDA is still crucial in making sensible primary and secondary assumptions. For example, suppose the analyst wants to fit some sort of regression model. Then a scatter diagram should indicate the shape of the curve (linear, quadratic or whatever) as well as give guidance on secondary assumptions (normality? homogeneous variance? etc.). Thus IDA is vital in selecting a sensible model and inhibiting the sort of 'crime' where, for example, a straight line is fitted to data which are clearly non-linear.

In a similar vein, a time-series analysis should start by plotting the observations against time to show up important features such as trend, seasonality, discontinuities and outliers. The time plot will normally be augmented by more technical tools such as correlograms and spectra, but is the first essential prerequisite to building a model.

As a third example, you may be interested in the interrelationships between a set of variables which arise 'on an equal footing' and are thinking of performing a factor analysis or a principal component analysis. If, however, you find that most of the correlations are close to zero, then there is no structure to explain and little point in such an analysis. On the other hand, if all the correlations are close to one, then all the variables are essentially 'measuring the same thing'. Then the main derived variable will be something like a simple average of the variables. I have frequently come across both situations with psychological data and have had to point out that the results of a formal multivariate analysis are not likely to be informative.

(h) Other things being equal (which they rarely are), prefer simple methods to complicated alternatives, while avoiding the simplistic.

The wrong technique is chosen more often than might be expected, not only by non-statisticians (Gore *et al.*, 1977), but also by statisticians. Our technique-oriented training helps us to know the theory and computational details of different techniques but does not help us choose the most appropriate method. Even when the right method is chosen, it could still be carried out incorrectly because of non-standard data or because we do not understand how to use the computer software available. We need more training in the use of software and the interpretation of computer output – which is often not as clear as it might be (Chapter 8).

7.4 Different approaches to inference

This book has hitherto adopted a rather pragmatic approach to the analysis of data, and has said little about the philosophical problems involved in assessing probabilities and making inferences. This section, which is rather different in spirit to the rest of the book, makes some brief remarks on these questions. As well as commenting briefly on different possible approaches to statistical inference, there are some general remarks on scientific modelling and on the history of statistics.

Statistical inference can be described as being concerned with making deductions from relevant information so as to obtain a description or model of the given situation (i.e. going from *knowns* to *unknowns*, or from the *particular* to the *general*). This includes the familiar problems of using a probability sample to make assertions about the population from which it was selected, and of using a given set of data to make deductions about the underlying probability model.

A first obvious question is: 'Why do we need a general formal theory of inference?' Some possible answers are that we need (i) a general framework for **teaching**, (ii) a general basis for **measuring uncertainty**, and (iii) guidance on strategy to follow in **applications**, especially when working from first principles with a new model in some unfamiliar setting.

It is not easy to give a taxonomy of different approaches to inference which will satisfy everyone. The two main approaches are the frequentist approach and the Bayesian approach. In the **frequentist** approach, the sample data are regarded as the main source of relevant information and the approach leans on a frequency-based view of probability in which the probability of an event is the proportion of times the event occurs 'in the long run'. There are some differences between the Fisherian approach which emphasizes the use of the likelihood function and the Neyman–Pearson approach which emphasizes operational considerations such as power, but they will not be discussed here. Note that **likelihood inference** is sometimes seen as a distinct approach. **Bayesian inference** combines **prior** information (the knowledge we have before seeing the data) with the sample data via Bayes' theorem so as to evaluate the **posterior** information. The approach leans on a subjective view of probability in which the probability of an event measures a person's degree of belief in the truth of some proposition or event based on the information held by that person at that time. There has been much attention to the case where there is little or no prior information (how do you represent prior ignorance?), and also to the (more interesting?) case of informative personalized priors. For subjective probabilities, it need not be necessary that the event of interest can be regarded as having been generated by some random process. A third general approach to inference is **decision theory** which aims to choose the 'best' decision from a prescribed list of possible actions using some sort of **utility** or **loss** function to specify the consequences of the different actions. Note that it can be difficult in practice to provide quantitative measures of the consequences of wrong actions, and that it is possible to formulate many frequentist and Bayesian inference problems as decision-theoretic problems. The latter emphasizes the blurred borderline between different approaches. Other approaches include **fiducial inference** (an attempt, which few people fully understand, to arrive at Bayesian-style conclusions without specifying priors) and **empirical Bayes methods** (a hybrid of the frequentist and Bayesian approaches in that data are used to estimate the prior).

Which approach to inference should the applied statistician adopt? If you look at the statistical literature, you will find strong disagreement between different writers as to what is the 'best' approach. If the sample is large, the qualitative conclusions resulting from the different approaches are likely to be similar (though the form of the interpretation may appear to differ). For small samples the situation is much trickier, though there will still be close agreement between the practical results of the frequentist approach and a Bayesian approach with 'prior ignorance'. Each approach has its ardent supporters but I am pleased to note that many statisticians (e.g. Box, 1994, p. 217) are adopting a flexible, pragmatic approach to statistical inference in which they refuse to 'label' themselves but rather see different approaches as being relevant to different situations, or else adopt a hybrid strategy – see below. The immense variety of statistical problems which can arise in practice and the flexible, interactive approach which is needed to solve them, make the long-standing arguments between different schools of inference seem rather academic and irrelevant to practical data analysis. A single approach is not viable (Cox, 1986) and we need more than one mode of reasoning to cope with the wide variety of real-life problems. In particular, *no general theory can easily take account of all contextual considerations* (Lehmann, 1993, p. 1247), such as those involved in formulating a model (section 5.3.1), while with messy data and unclear objectives, the problem is not how to get the *optimal* solution, but how to get *any* defensible solution. In the latter type of situation, philosophical problems seem relatively unimportant. Moreover, most practical problems involve some sort of informal data-analytic activity (e.g. EDA or IDA) which again makes it difficult to rely on any single formal theory of inference.

Many statisticians adopt a frequentist approach for routine analyses, perhaps more for computational convenience than because the underlying rationale is necessarily thought to be convincing. For example, statisticians know that the P-value resulting from a significance test is often not what the analyst really requires and that it is often misinterpreted (section 7.2), but P-values are still widely computed, partly because of 'public demand', partly because they do provide some, albeit limited, evidence, and partly because alternative approaches also have their difficulties.

The Bayesian approach has much to commend it as a theoretically sound, coherent approach to formal inference. However, it requires the analyst to represent prior information in a particular, very formal, way, and provide prior probabilities for models or hypotheses and prior probability distributions for unknown model parameters. This can be difficult in many practical problems and raises many contentious issues. Scientists may not be very good at quantifying their prior information (which may be nothing more than intuitive prejudices) and many statisticians prefer to restrict attention to the data and incorporate judgemental issues in an informal qualitative way or leave such judgements to others. Furthermore, Bayesian computations are typically much heavier than those for the frequentist approach, although modern computing software is slowly overcoming this problem. If a rigid Bayesian viewpoint is rejected (as I think it should be), it is still worth recognizing the insight given by the approach, and the superiority of Bayes to frequentist solutions for some types of problem. Frequentists can learn much from the Bayesian approach (and vice versa), and some statisticians adopt a hybrid of the two. For example, the long-run frequency interpretation of a confidence interval is unappealing but many statisticians are not afraid to interpret such frequentist probabilities in a subjective way where this seems appropriate. Although theoretically dubious, this often helps in practice.

In fact the philosophical and practical arguments between advocates of different inferential approaches, fascinating as they may be, have mainly concerned just one part of the statistical process, namely the fitting of an *assumed* model to a single set of data. In practice, 'the choice of model is usually a more critical issue than the differences between the results of various schools of formal inference' (Cox, 1981). This book has argued that we should see statistical inference in a wider context by recognizing three different stages of inference, namely (i) model formulation, (ii) model fitting; (iii) model checking. Furthermore, in Chapters 2 and 5, we noted that in practice there may be several cycles of model fitting as defects in some original model are recognized, more data are collected, and the model is gradually improved. We also noted that there is no such thing as a true model, but rather that some models are useful while others are not. The iterative nature of model building (and of learning in general), and the fiction of a 'true' model, are further reasons why the principles of model formulation and checking are more important than the disagreements between the different schools of inference, especially in their more sterile manifestations. Statisticians need to get more involved with modelling issues, and more general guidance is badly needed. However 'good' the statistical inference may be, it is still conditional on the assumed model. Poor models can lead to silly answers and to time and money being wasted (though disasters are usually hushed up!).

Inference also tends to concentrate on the estimation of **unobservable** parameters rather than on the use to which a model will be put, such as the prediction of observables (which allows the model to be **calibrated** – see section 5.3.4).

The different approaches to inference also fail to emphasize the distinction between looking at a brand new set of data and looking at a series of similar data sets. The problems involved in taking several samples and in combining information from different sources have received relatively little attention from statisticians. Draper *et al.* (1992) review existing work, much of it of an interdisciplinary nature including meta-analysis (section 5.2), the use of hierarchical models, and the combination of P-values from different studies (definitely not recommended). Ehrenberg's (1982; 1984) approach to data analysis and model building emphasizes the desirability of establishing regular patterns across several samples of data, describing these patterns in a suitable way, and finding ways of highlighting departures from the model. A 'first-time' analysis needs to be followed up with further studies so that knowledge is built up through empirical generalization, leading to a model which Ehrenberg calls a 'lawlike relationship'. This model may, or may not, be stochastic in nature. Of course, no two studies can be made under exactly identical conditions. In any case, rather than try to replicate results exactly (though this is sometimes desirable). Ehrenberg and his colleagues (Ehrenberg and Bound, 1993; Lindsay and Ehrenberg, 1993) argue that replications should be highly differentiated so as to show which factors do, or do not, affect the results. The critical issue is to find a model which holds, at least approximately, across many data sets (as well as satisfying external requirements – see section 5.3.1) rather than one which happens to fit one set of data 'best' (though it may perhaps be only marginally superior to alternative models).

When a model is fitted to many data sets, it is much more important to assess the overall agreement than to question the fine details of a particular fit, though systematic local deviations will be of interest. The point is that prior knowledge about empirical regularities should be used, when available, to prevent the analyst 'reinventing the wheel' every time a new set of data is acquired. For example, section 5.3.3 refers to a general model of consumer purchasing behaviour which I have helped to construct using numerous data sets. Methods for analysing a series of data sets have not received the attention they deserve, partly because they do not lend themselves to theoretical (i.e. mathematical) analysis, and partly because many statisticians do spend most of their time analysing more or less unique data sets. The latter state of affairs is arguably rather unbalanced. If statisticians wish to be regarded as fully-fledged scientists (see below) then they need to be able to cope with both brand new and continuing sets of data.

7.4.1 Statistics and science

This section gives a very brief overview of science and its relationship with statistics. It is arguable that a full appreciation of statistical inference can only come through a careful study of the nature of scientific modelling (e.g. Penrose, 1989; Box, 1976; 1990). Science is concerned with deriving general laws from particular experiments. The scientist does not normally rely on a single experiment but prefers to collect lots of data. Moreover, the scientist often wants to predict *outside* the population from which, strictly speaking, earlier data have been sampled. For example, one hopes and expects that Boyle's law will hold not only for different gases last week, but also next week, in different countries, in different laboratories, and so on. Such a result can only be established by taking many samples under a variety of external conditions.

So, how does statistical inference relate to scientific inference? Questions like this have been asked ever since the formation of the Royal Statistical Society 150 years ago, but no clear answer has emerged. Box (1990) has said that 'Statistics is, or should be, about scientific investigation and how to do it better', and has also said that statisticians should *strive to become first-class scientists rather than second-class mathematicians.* I heartily agree with these sentiments.

Statistics is often called statistical science and is therefore presumably widely regarded as being a science. But is it? A knowledge of statistics is certainly vital in collecting 'good' data and in making inferences, but in practice, as noted above, statistics is often (too often?) concerned with 'one-shot' experiments rather than the more general relationships usually sought in scientific invest-igations. The latter typically follow an iterative sequence involving more than one data set, but the questions this raises are not addressed in much inferential theory.

Another important question which affects statistical inference, is whether statistics is **inductive** or **deductive**. At first sight, it probably appears inherently inductive (data → model) but the overall scientific process involves an iterative inductive–deductive cycle. Karl Popper and others have pointed out the importance of inspiration, speculation, experience and even guesswork in model building and this type of process cannot readily be examined mathematically. Popper asserted that scientific theories are not derived inductively from the data, but rather as speculations, and are then subjected to experimental tests in an attempt to refute them. A theory is entitled to be considered scientific only if it is in principle capable of being tested and refuted. When a theory survives such tests, it acquires some credibility and may eventually be regarded as established. But it can never be *proved* true. Popper's views are somewhat extreme but they have done away with the idea that science is entirely inductive. But neither is it entirely deductive. Thus it is unfortunate that statisticians too often prefer to concentrate on the deductive part of the process which can be studied theoretically and pretend the other components do not exist. Then students get the impression that 'models are true and data are wrong instead of the other way round' (Box, 1990). Given the importance of subjective judgement and inspiration in the modelling process, it follows that neither science nor statistics is fully objective (despite the fact that people expect them to be). Choices have to be made. Even in the 'one-shot' case, statistical inference is not 'objective' in that it demands, for example, an understanding of the context (e.g. which variables to measure in the first place), exact knowledge of the measurement process, and a clear appreciation of the problem which needs to be solved. Instead of arguing so much about the different approaches to inference in the 'one-shot' case, it should prove more rewarding to work towards establishing a more unified view of the sciences as a whole and of the role for statistics within this.

7.4.2 A brief history of statistics

In order to set the above remarks in context, it may be helpful to close this section with a (very) brief review of the history of statistics. We can often learn much about a subject by studying its history.

Modern **probability theory** started around 1700, while the use of least squares to combine information started around 1800. Even so, before about 1900, statistics was mainly restricted to what is now called descriptive statistics. Despite (or perhaps because of!) this limitation, statistics made useful contributions in many areas. **Censuses**, which may be regarded as the first quasi-statistical activity, were taken for the purpose of taxation as early as 3000 BC in ancient Babylon, China, Egypt and later in the Roman Empire. They started in the Western world in the 18th century and have, for example, been carried out at regular intervals (usually every 10 years) since 1790 in the USA and 1801 in Great Britain.

The routine recording of other statistical information grew steadily over the years. For example, Florence Nightingale is most famous for her role as a nurse in the Crimean War, but she may have had more lasting influence as a nursing administrator and as the first female Fellow of the Royal Statistical Society. Around 1853, she embarked on a simple, but effective, programme for collecting data on hospital admissions and fatalities and noticed, for example, that women giving birth in hospital were much more likely to die than those giving birth at home. This led eventually to the establishment of separate wards for mothers-to-be so that the chances of cross-infection were reduced. This may seem obvious today but was not then! After her return from the Crimean War (1856), she used statistics to promote a variety of reforms, first of health in the British Army, and then of public health generally. All this despite her own failing health. 'A superb propagandist, her development of ways of expressing statistical data through diagrams brought statistics to the attention of the general public in a new and immediate form' (Diamond and Stone, 1981). Nowadays the managerial and administrative demands for more and more data may have swung the pendulum too far the other way so that some health workers spend unreasonable lengths of time recording information, perhaps even to the detriment of patient care. This is sad.

The progression from just recording data to collecting experimental and survey data was slow. As early as 1747, a simple **experiment** on board ship showed that scurvy patients who were given citrus fruit were cured in a few days and were able to help nurse other scurvy patients. In 1835 a Frenchman, Pierre Louis, showed that blood-letting was harmful rather than beneficial to patients. More fundamentally, he argued that all therapies should be open to scientific evaluation. It may seem incredible that blood-letting could have been used so widely without supporting evidence, but how many treatments in use today are based on medical folklore rather than hard fact? For example, one expensive drug, with rather nasty side-effects, was enthusiastically prescribed by doctors in the 1950s for a certain condition (retrolental fibroplasia) because it appeared to give a 75% 'cure' rate. However, a proper clinical trial showed that 75% of patients were 'cured' with no treatment whatsoever and so the use of the drug was discontinued. Of course the conduct of clinical trials is often complicated by the ethics of prescribing a particular treatment (which might be no treatment at all) when the body of belief is that an alternative treatment is better, no matter where or how this belief originated.

An interesting account of the evolution of **sample surveys** in Great Britain is given by Moser and Kalton (1971, Chapter 1). The classical poverty surveys of Booth and Rowntree about the turn of the century were instrumental in alerting people to the terrible conditions widely prevalent at that time, while the first major use of carefully conducted sampling was probably that of Bowley in 1912.

As the 20th century advanced, the use of statistics in alliance with scientific experimentation increased steadily. This generated the need for a more formal inferential apparatus to analyse data sets which were often quite small in size. One thinks particularly of the early biometrical school of Francis Galton and Karl Pearson and the new biometrical school of Sir Ronald Fisher. The latter's work was particularly influential in laying the foundations for much of modern statistics, including classical inference and experimental design, and it is noteworthy that much of this work was carried out while Fisher was working at Rothamsted Agricultural Research Station in the 1920s. Many of the most important advances in statistics have been made by people working 'in the field'. Fisher's inferential ideas were not without their critics, notably Karl Pearson and Jerzy Neyman, and there were some quite spirited (some might say vicious) debates between protagonists of different views.

The Bayesian approach has developed mainly since the Second World War, although the original papers by the Reverend Thomas Bayes on inverse probability were published some 200 years ago. In the last few years, advances in computation have meant that the approach can at last tackle more practical problems, but I find it disquieting that some Bayesians can be so critical of statisticians adopting alternative approaches, not only of frequentists, but even of fellow Bayesians who do not rigidly follow the 'party' line. I look forward to a more tolerant intellectual environment which will work to the benefit of all.

By the 1960s, the pendulum had swung from descriptive statistics right over to mathematical statistics, with the result that many textbooks neglected the practical aspects of analysing data while published journals became very theoretical. Practising statisticians began to question this balance and Moser (1980) described some published research as being 'theories looking for data rather than real problems needing theoretical treatment'. Since the 1980s the pendulum has rather swung back with much more emphasis on data analysis. This has been helped by improvements in computing facilities leading to substantial advances in computational aspects of statistics, including such topics as interactive graphics, image analysis and bootstrapping. The resurgence of data analysis has also been helped by the publication of various relevant books and papers (see the references listed in section 6.9.2, especially Tukey, 1977). One aim of this book is to continue this revival through increased emphasis on IDA.

Barnett (1982, p. 308) suggested that descriptive data-analytic methods can almost be regarded as yet another general approach to inference. However, I have argued that IDA is useful in both data description and model formulation, and so I would prefer to see data-analytic methods recognized as an essential ingredient of a broad composite form of inference which does not have a narrow philosophical base, but which allows the analyst to adopt whatever procedure appears to be appropriate for a particular problem. This ecumenical (e.g. Box, 1983) or eclectic approach to statistics is one that I hope and expect to mature over the coming years.

Further reading

Barnett (1982) gives a comprehensive account of comparative statistical inference. Oakes (1986, Chapter 6) gives a critique of all the main approaches to inferences and concludes that 'none of the approaches is entirely satisfactory'. Bernardo and Smith (1994, Appendix B) give a critique of non-Bayesian approaches. There is much historical information in the entertaining book by Peters (1987), while Stigler (1986) gives a more detailed account up to 1900.

8
Using resources — 1: The computer

The vast majority of statistical analyses (outside the Third World) are now carried out using a computer. Thus the statistician needs to:

1. understand the important features of a *computing system* (this should be an integrated combination of hardware and software);
2. know at least one scientific programming language;
3. know how to use some of the most popular packages;
4. be able to construct and edit a data file.

The choice of computer and its accompanying software, and its subsequent use in solving problems, is clearly crucial. Unfortunately, there is often little guidance on statistical computing in statistical textbooks. This is partly because the range of available material is so bewilderingly wide, partly because of doubts as to what is relevant in a statistics book, and partly because the scene is changing so rapidly. New computers and new software arrive continually, and computing power is still growing rapidly. Today's desktop microcomputer is more powerful than the mainframe computer of a few years ago. Thus this chapter concentrates on general remarks which I hope will not go out of date too quickly.

A computer enables much arithmetic to be performed quickly and accurately. It allows the analyst to look at data in different ways, try a variety of graphical devices, and fit (many) models in an iterative, interactive way. On the debit side, the wide availability of computer software has tempted some analysts to rush into using inappropriate techniques. Unfortunately, most computer software is not yet intelligent enough to stop the user doing something stupid. The old adage 'garbage in, garbage out' still holds good, and it must be realized that careful thought and close inspection of the data are vital preliminaries to complicated computer analyses. Another potential drawback to using a computer is that it allows **data mining**, whereby so many models are tried that a spuriously good fit is obtained (section 5.3).

Despite the increasing use of computers, many students still spend considerable time working through problems by hand or with a pocket calculator. While this can sometimes be helpful to understand a method thoroughly, I suggest that students who do have access to a computer need more help and training in learning how to interpret computer output. As well as getting used to 'reading' routine output, they need to be prepared for unexpected and incomprehensible messages. For example, some programs routinely print a statistic called the Durbin–Watson statistic, without saying what it is or what a 'normal' value should be. Many users have no idea what it means!

The computer not only makes existing statistical techniques easier to use, but also has growing influence in generating new innovative techniques. Research topics of strong current interest include computationally based inference methods (especially **resampling** methods – see Appendix A.4), **smoothing** methods, **pattern recognition**, **image analysis** and a variety of computational algorithms. Such methods raise many new, interesting problems, not least of which is the general question as to how computer-generated solutions can be checked for correctness.

Computers have also led to increasing use of **automatic data-recording**, for example with medical monitoring machines (see also section 6.3). This can be very efficient but also requires careful monitoring. If the data are also analysed automatically using some sort of 'black box', then things can go badly wrong in many ways. The user must be satisfied that the 'black box' is working correctly and should in any case check that the results 'look reasonable'.

When choosing a personal computer, desirable features include versatility, convenient data input and data storage facilities, high-quality graphics, and a wide range of good software. However, many statisticians will have to use an in-house computing system and this in turn specifies some aspects of the man–machine interface, such as the operating system. Faced with a specific problem requiring computation, the analyst must decide whether to use a computer package, to augment a published algorithm, or to write a special program, which may be a one-off program or a more general program which could be used with similar subsequent sets of data.

If writing a program, include plenty of comment statements, especially if other people are going to use it, and test different combinations of input variables. It is helpful to write the program in modules which can be tested separately. The choice of language usually depends on a variety of in-house considerations such as the computer, compatibility with other programs, and portability.

As regards algorithms, the reader should realize that good ones have been published to cover the vast majority of mathematical and statistical operations. It is bad practice to 'reinvent the wheel' by trying to write a set of instructions from scratch when a published algorithm could be used. For example, algorithms are printed in journals such as *Applied Statistics* and the *Computer Journal*, and there is a comprehensive range of algorithms published in the USA by the Institute of Mathematical Statistics (IMSL routines) and in the UK by the Numerical Algorithms Group (NAG routines). These cover such topics as interpolation, curve fitting, calculation of eigenvalues, matrix inversion, regression, ANOVA, and random number generation.

The need to write special programs, or even use algorithms, is slowly diminishing as a wider and wider range of software becomes available, and a separate section will be devoted to the latter topic.

8.1 Choosing a software package

Packages vary widely both in quality and in what they will do, and the choice between them may not be easy. Some have been written by expert statisticians, but others have not. Some print out plenty of warning, error and help messages, but others do not, and may indeed go 'happily' on producing meaningless results. Some are written for one small area of methodology while others are more general. Some are written for expert users, while others are intended for statistical novices. Unfortunately, while packages make techniques much easier to implement, they also make them easier to misuse and this has led to widespread abuse by non-statisticians. We need packages which incline towards expert systems (section 8.2) where the package will be able to say, for example, that a given set of data is unsuitable for fitting with such-and-such a model.

A few packages still run in batch mode, where the analyst has to decide beforehand exactly what analyses are to be carried out, but most packages now run in interactive mode, where the user can react to interim results. The command structure of an interactive package may be what is called 'command-driven' or 'menu-driven'. For the latter, a range of options is given to the user at each stage from which one is selected. This type of system is more suitable for the inexperienced user. Some packages allow completely automatic analyses of data, where the analyst abdicates all responsibility to the computer, but interactive analyses are usually preferable.

Software needs to be appraised on various criteria which include statistical, computational and commercial considerations. Desirable features include:

- flexible data entry and editing facilities;
- good facilities for exploring data via summary statistics and graphs;
- statistically sound procedures for fitting models, including diagnostic checking;
- computationally efficient programs;
- well-designed, clear and self-explanatory output – unambiguous estimates of standard errors should be given and excessive numbers of significant digits avoided, and there should be more emphasis on the estimation of effects and less on P-values;
- adequate documentation and support.

Other criteria include the cost of the package, how easy it is to learn and use, the required equipment, the needs of the target user and the possibility of extending the package.

Regarding computational aspects, I note that many packages were written when computing power was a major constraint. Nowadays the accuracy of algorithms is probably more important than speed and efficiency. It is difficult to assess numerical accuracy and efficiency except by running specially selected, and perhaps unusual, data sets. Although computing power is now enormous, it must be realized that numerical 'bugs' are still alive and 'kicking' and so users must beware, particularly when trying a new package. A set of test (or benchmark) data, where the answers are known, should always be run to see if a package can be trusted. To ensure statistical validity, it is important that professional statisticians fight to retain control of the development of statistical computing.

As to documentation, manuals should be easy to read, include examples, and allow easy access for specific queries. Unfortunately, this is often not the case and separate handbooks have been written for some packages to clarify the manuals. Established packages are often better supported, maintained and updated than some newer packages. Good packages will also have plenty of online help so that the user rarely needs to look up the manual anyway.

The needs of the target user may vary widely, and one can distinguish the following categories:

1. the expert statistician who may want to use established methodology in consulting or collaborative work, or carry out research into new methodology;
2. teachers and students of statistics;
3. the statistical novice who understands little statistics but knows what he/she wants to do (even though it may be quite inappropriate!).

Ideally a package should be *robust to user variation* and be useful for all the above categories, but in practice it is hard to design software to meet the needs of all these potential customers at the same time.

8.1.1 Examples of useful packages

New and revised packages for computers and microcomputers are being released continually and it is impossible to provide an up-to-date review of them all here. This section concentrates on well-established packages, written originally for mainframe computers, which have mostly become available for microcomputers as well. Note that most packages are now available in a **Windows** version which is generally easier to use. Also note that the comments made below at the time of writing may become outdated as packages are improved. The reader should keep up to date by referring to reviews of computer software in various journals such as the *American Statistician* and selected computing journals. Your computer unit should also be able to provide further information including manuals for any software which is already available locally.

As new versions of packages are constantly appearing, the following remarks are mostly of a general nature. When presenting computational results, it is advisable to cite software in full including the exact version of the software and perhaps even the make and configuration of the machine. Since different packages use different algorithms even for common procedures, this is the only way to ensure that results really are replicable. For example, different packages use different estimation routines to fit time-series autoregressive models and the resulting differences are not trivial for small sample sizes. Even worse is the package which calculates a standard deviation by dividing by the sample size, n, in one routine, and by $n - 1$ in another!

MINITAB is an interactive, command-driven package which covers such topics as exploratory data analysis, significance tests, regression and time-series analysis. It is very easy to use and is widely employed by both commercial and academic institutions. At my own university, we use if for teaching both introductory and intermediate courses. The expert will find it too restrictive for some purposes. A brief summary is given in Appendix B.1. The book by Ryan and Joiner (1994) may be preferred to the package manual.

GENSTAT is a statistical programming language which allows the user to write programs for a wide variety of purposes. It is particularly useful for the analysis of designed experiments, the fitting of linear models (including regression) and generalized linear models. It also covers most multivariate techniques, time-series analysis and optimization. It allows much flexibility in inputting and manipulating data. Programs can be constructed from macros, which are blocks of statements for a specific task. A library of macros is supplied. The package is designed for the expert statistician and is not user-friendly. It can take several days to learn and so needs to be used regularly to make it worthwhile. Snell and Simpson (1991) describe a series of examples using the package.

BMDP-77 is a suite of programs which covers most statistical analyses from simple data display to multivariate analysis,. BMD stands for biomedical, and the programs were written by a group at UCLA (University of California at Los Angeles). Snell (1987) describes a series of examples using the package. This comprehensive package is good for the professional statistician and has good user support, but can be rather difficult to learn.

SPSS denotes 'statistical package for the social sciences'. This comprehensive package is probably used more than any other package, but mainly by non-expert users. Some statisticians view the package with some reserve. It produces 'answers' of a sort even when the 'question' is silly, can be tricky to use, particularly at first, and there is concern that it is widely misused by the non-expert user.

GLIM denotes 'generalized linear interactive modelling'. It is an interactive, command-driven package which is primarily concerned with fitting generalized linear models. This means that it covers regression, ANOVA, probit and logit analysis and log-linear models. The user must specify the error distribution, the form of the linear predictor and the link function. The package is powerful, but requires considerable statistical expertise, is not user-friendly, and can produce output which is difficult to interpret (e.g. estimates of main effects are given in a non-standard form). Nevertheless, I do sometimes use it to fit generalized linear models as it can do things which are difficult to perform using other packages (except GENSTAT). It is relatively inexpensive. A brief summary is given in Appendix B.2. Note that everything which can be done using Release 4 can also be done using Release 2 of GENSTAT 5, albeit using a different (more complicated?) syntax.

S-PLUS is an interactive language designed for the expert statistician which covers a wide range of procedures. It is particularly good for graphics.

SAS denotes 'statistical analysis system'. This programming language is widely used, both in the USA and the UK, particularly in pharmaceutical applications. It is suitable for the more advanced statistician. I have heard good reports of the package but have no personal experience of it. It is rather expensive compared with other packages, and is not the sort of package which can be learnt in a day.

There are numerous other general and more specialized packages available and they cannot all be covered here. STATGRAPHICS is a general starter package which seems to be easy to use and generally well liked. HARVARD GRAPHICS and FREELANCE are two useful specialist graphics packages. Packages in my own special-interest area of time-series analysis include AUTO-CAST for Holt–Winters forecasting, AUTOBOX for Box–Jenkins forecasting, BATS for the Bayesian analysis of time series and many more. There are literally hundreds of packages and the intending buyer should shop around.

There are a number of packages which are much more mathematically oriented. They include MATLAB, MAPLE and MATHEMATICA. As well as some data-analysis procedures, they enable the user to work with a variety of mathematical functions, carry out numerical analysis and matrix algebra, plot 2-D and 3-D functions and graphs, and much, much more.

One class of packages not mentioned so far are **spreadsheet** packages. They allow the easy manipulation of data in the form of a two- or three-way table of numbers and also allow some simple statistical analyses to be carried out on them. They are widely used, and the three best-known packages are LOTUS 1-2-3, EXCEL and QUATTRO PRO.

The final package to be mentioned here, called DBMSCOPY, has a completely different purpose, namely to enable data to be transported from one package to another when they are not in compatible format. The question of portability of data between packages and computers is increasingly important.

8.2 Expert systems

This section discusses a general type of package called an **expert system**, which aims to mimic the interaction between a user of statistics and a statistical consultant. A good system will do some, or even all, of the following:

1. help make the right decision;
2. help refine the research question by asking questions;
3. help choose an appropriate technique;
4. apply the technique;
5. provide information;
6. offer intelligent advice;
7. interpret the results.

While it is difficult to define an expert system exactly, it should certainly be distinguished from an ordinary package, however good, and also from a knowledge-enhancement system, which simply aims to provide information. Expert systems attempt to incorporate the sort of questions and advice that would come from an experienced statistician in regard to clarifying objectives, exploring data, formulating a model and choosing a suitable method of analysis. Ideally, they should also offer some explanation of the argument that has been used to come to any particular decisions made. Expert systems have already been introduced in areas such as medicine, but there are still rather few statistical packages available that can really be described as expert systems, and further research is needed to implement them more widely. In particular, the writing of a program to mimic statistical consulting requires additional research into one of the main topics of this book, namely the *strategy* of problem solving.

The development of 'good' expert systems could also have useful side-benefits in regard to ordinary packages. Many of the latter are currently unfriendly, or even dangerous, for the non-expert user and it seems likely that some expert system features could be routinely included in more packages. For example, the user of an interactive package could easily be asked questions such as 'Are you sure that the following assumptions are reasonable?' followed by a listing of the assumptions implicit in the particular method being used. In this way packages would move towards being expert systems.

For further reading, I suggest the introduction in Hand (1994) and the collections of papers in Hand (1993) and in the *Journal of Applied Statistics*, Volume 18, No. 1, 1991.

9
Using resources — 2: The library

Statisticians, like other professionals, cannot be expected to 'know everything'. However, they must know how to locate appropriate reference material, when necessary, and be able to understand it when found. A library is the most important source of knowledge, and, used wisely, can be a valuable aid in tackling statistical problems.

Libraries contain a richer variety of material than is sometimes realized. As well as books, they usually contain a range of statistical journals, various abstract and index journals as well as tables of official statistics (see Exercise G.6).

9.1 Books

Obviously libraries contain a range of textbooks and reference books. Some of the more important ones are listed in the references at the end of this book, and they can give valuable help when tackling difficult or unfamiliar problems.

In order to find a book on a given topic, the book index may be used to look up a specific author, a specific title, or a keyword such as 'forecasting' which forms part of the title of a book. In addition, it may be worth searching all books with a given code number, such as those coded under 'time-series analysis'.

9.2 Journals

Recent research in statistics includes developments in both the theory and practice of statistical methods as well as new reported applications in specific problem areas. This research is usually reported in one of the many statistical journals before finding its way into book form at a later date. It is worth having a 'browse' through these journals to see what sort of material is available. The most important journals are as follows: The (British) Royal Statistical Society publishes four journals:

1. Series A – concerned with statistics in society, and has good book reviews;
2. Series B – concerned with the theory of statistics;
3. *Applied Statistics* (Series C) – self-explanatory;
4. *The Statistician* (Series D) – general articles (formerly published by the Institute of Statisticians, now merged with the Royal Statistical Society).

The American Statistical Association publishes five journals:

5. *Journal of the American Statistical Association* – a mixture of theory, applications and book reviews;
6. *American Statistician* – a readable quarterly journal including tutorial articles and reviews of computing software;
7. *Technometrics* – published jointly with the American Society for Quality Control and concerned with the development and use of statistics in science and engineering;
8. *Journal of Business and Economic Statistics* – self-explanatory;
9. *Journal of Computational and Graphical Statistics* – published jointly with the Institute of Mathematical Statistics and Interface.

The International Statistical Institute publishes:

10. *International Statistical Review* – mainly review articles.

The Institute of Mathematical Statistics publishes:

11. *Statistical Science* – readable review papers.

Other journals include:

12. *Biometrics* – with applications in the biological sciences, published by the Biometric Society;
13. *Biometrika* – mainly theory.

There are also numerous more specialized journals such as:

14. *Statistics in Medicine*;
15. *Journal of Marketing Research*;
16. *International Journal of Forecasting*.

Statistical applications (and occasionally new methodology) also appear in many non-statistical journals, but the general quality is very variable.

With thousands of articles published each year, it is impossible to keep up with all statistical developments. The statistician must be judicious in choosing which journals to scan, and even more selective in deciding what is actually worth reading. As an academic, I look at all the above journals, but some readers may wish to confine attention to, say, journals 1, 3, 5 and 7 above.

Abstract and **index journals** complement and augment the above journals. These specialized journals do not contain written articles, but rather contain lists or brief summaries of articles in other journals. For example, *Statistical*

Theory and Methods Abstracts contains brief (e.g. half-page) summaries of papers in statistical journals. However, I have found the index journals more useful, mainly for finding papers on particular topics. The *Science Citation Index* contains a list, alphabetically by author, of all papers published in science journals (which includes many statistical journals). It actually consists of three journals, called the *Citation Index*, the *Source Index* and the *Permuterm Index*. Suppose you want to see if anyone has followed up the work of Dr X.Y.Z. Smith in 1970. You look up X.Y.Z. Smith in the *Citation Index* which lists all recent papers which refer to the earlier paper. Full details of these recent papers may be found in the *Source Index*. The *Permuterm Index* enables you to look up keywords taken from the titles of recent papers. There is a separate *Social Science Citation Index*. There is also a *Current Index of Statistics*, which is solely concerned with statistical journals and enables the user to find, for example, all papers whose title includes the word 'forecasting'.

9.3 Other sources of statistical information

Most countries have a national statistical office which issues official statistics in various publications at various intervals of time. Such statistics include demographic statistics (birth, deaths, etc.), economic statistics (e.g. cost of living, number of unemployed), social statistics (leisure, etc.), as well as statistics on crime, housing, education, etc. In the UK, for example, the Central Statistical Office publishes a monthly digest and annual abstract of the more important statistics, the annual *Social Trends*, aimed at a general audience, as well as a variety of more specialized publications such as *Economic Trends*. They also publish a *Guide to Official Statistics* (in its fifth edition in 1990) and an annual booklet entitled *Government Statistics – a brief guide to sources*. In the USA, the US Bureau of the Census publishes a wide range of statistics. Various regional, foreign and international statistics are also usually available and the *Demographic Yearbook* and *Statistical Yearbook* published by the United Nations are particularly helpful. The range of publications is so wide that there are specialized books giving guidance on sources of statistics.

10
Communication — 1: Effective statistical consulting

Many statistical projects arise from requests for help from specialists in other disciplines. Such specialists will have varying degrees of expertise in statistics, and the ability to communicate effectively with them is most important. Such work is usually called **consulting**, although Cox (1981) has questioned the overtones to this word and advocated greater emphasis on **collaboration**. It is certainly true that full participation in a collaborative study is more rewarding than trying to give a quick answer to a cookbook question. However, in my experience, the statistician must be prepared to give advice at a variety of levels and the following remarks are concerned with statistical consulting in its widest sense.

The first positive remark to make is that consulting can be constructive, stimulating and intellectually satisfying, as well as fun! I (usually) enjoy it. Against this, it has to be admitted that there are many reasons why consulting or collaboration may sometimes 'break down' and both the statistician and client may be at fault. Instead of seeking impartial statistical advice, the client may simply want the statistician to do his work (or 'number crunching') for him. Alternatively, the client may simply want a publishable (i.e. 'significant') P-value, the confirmation of conclusions which have already been drawn, or some 'miracle' analysis if he realizes the data are of poor quality. Sometimes clients get what they ask for, even if it is not what is really needed, but sometimes they get more than they originally asked for if the statistician's probing questions lead to a reformulation of the problem. Unfortunately, the client may get frustrated because the statistician does not understand the problem, or insists on solving a different one, or comes up with unintelligible conclusions, or takes the data away and is never seen again! As regards the last point, it is difficult to handle the conflicting pressures to *be thorough* and yet *report on time*. It is therefore good practice to impose strict deadlines on practical projects for students so that they learn to use time profitably and efficiently.

Much of the following advice may appear 'obvious' common sense but is not always followed in practice. The statistician should:

1. Have a genuine desire to understand and solve real problems and be interested in the field of application.
2. Establish a good working relationship with the 'client' or collaborator so as to 'know the customer' and meet his/her needs.
3. Be prepared to **ask lots of probing questions**. There is a fine distinction between getting enough background material to understand the problem, and getting bogged down in unnecessary detail, but it is generally better to probe too deeply than not deeply enough. The importance of getting background information is illustrated in Example 5 of Chatfield (1991), while Examples 10.1–10.3 below also illustrate the importance of asking questions. Sometimes simple enquiries about apparently minor details can produce startling revelations or bring misunderstandings to light. It is unwise to take anything for granted. Some questions which are often useful are:

   - What is the background to the study?
   - What do you expect to learn from the investigation?
   - How will you use the results?
   - Have you worked on similar problems before?

   This initial question-and-answer session should enable you to ascertain what prior information is available. Be prepared to interrupt (politely, but firmly) if you do not understand what the client is saying, particularly if specialist jargon is used. Equally, you should try to avoid statistical jargon yourself. Handling a consulting session requires some organizational skill.
4. As part of this initial question-and-answer session, get the client to be precise about the **objectives** of the study (Chapter 3). Sometimes the real objectives turn out to be quite different from the ones first stated. Instead of asking 'What is the problem?', it may be better to say something like 'Tell me the full story', so as to find out what the problem really is.
5. Try to get involved at the planning stage so that the project becomes a **collaborative study** rather than a potentially unsatisfactory consultation. It is much harder (and perhaps impossible) to help if data have been collected without following basic statistical principles such as randomization. Regrettably, statisticians are often consulted too late. If data have already been collected, find out exactly how this was done and always ask to see the actual numbers. If the data are of poor quality, refuse to **dredge** them by applying overly sophisticated methods.
6. Bear in mind that **resource constraints** can play a large role in determining the practical solution. The statistician needs the skill to negotiate appropriate time and money to carry out a study properly, and perhaps also to negotiate for co-authorship of any resulting reports or papers.
7. Keep the design and analysis as simple as possible, consistent with getting the job done in the required time. Be willing to settle for a 'reasonably correct' approximate solution. And a partial solution to a problem is better than no answer at all!

8. Be prepared to admit that you cannot answer some problems straight away and may need time to think about the problem and consult another statistician or look things up in the library.

9. Refuse to answer questions over the telephone as this does not give you the opportunity to see the data or give you time to get sufficient background information.

10. Present the statistical conclusions and recommendations in clear language, preferably so that they can be understood by a layman.

Example 10.1 Asking questions

A consulting session will sometimes involve a series of questions along the following (rather frustrating) lines:

Q: What do you want to do with your data?
A: A *t*-test.

Q: No, I mean why have you carried out this study and collected these data?
A: Because I want to do a *t*-test (with puzzled look).

Q: No, I mean what is your prior hypothesis? Why do you want to do a *t*-test?
A: My supervisor/employer wants me to.

Q: No, I mean what is your objective in analysing these data?
A: To see if the results are significant. ...

It can sometimes be quite difficult to make progress! Even so, the statistician must persevere.

Example 10.2 Dealing with the scientist who 'knows what he wants'

(i) It is still important to ask questions and ask to see the data when dealing with researchers who apparently know what they want. For example, a colleague from our Chemistry Department asked me for two minutes of my time to answer a 'simple' question on regression. He had fitted a straight line to some data and wanted to know the formula for calculating the standard error of the estimated intercept. His knowledge of regression was clearly appreciable, and I could have just given him the formula. However, I asked to see the data first as I always do. The scatter plot showed clearly that the relationship was non-linear rather than linear, but my colleague calmly informed me that 'chemists always fit straight lines to this sort of data'. I pointed out that this was unwise and that if the wrong model was being fitted, the estimate of the intercept would be biased and the formula for its standard error would be inappropriate.

(ii) A colleague from the Biology Department told me that he had carried out a study to examine the fitness of a sample of 100 students from the university by getting them to run a mile at a time convenient to them and record the time taken. Eleven students failed to complete the course or return a time, but the remaining 89 students had running times whose mean and standard deviation were 6 minutes 15 seconds and 55 seconds, respectively. 'Would you test the null hypothesis that the average running time is 6 minutes? It is a t-test isn't it?'

This is a very typical question where it can be easier to do what is asked even when, as here, it is obviously (?) silly. You may not get much thanks for asking questions, clarifying objectives and querying the suggested approach, but an ethical consultant statistician should still do it. So what questions would you ask?

The first question might be whether the 100 students were selected randomly, and, if so, from what population. Are the 11 students who did not return a time missing at random, or are they the most unfit students? If the latter, then the data are effectively censored and the sample mean is misleading. And why is the given null hypothesis of particular interest? It sounds as though 6 minutes is an arbitrary time (which indeed it was). It turned out to be far more important to summarize the data in a sensible way than to carry out the (very artificial) significance test which was requested.

Example 10.3 Answering a question with a question

A good exercise for teaching the importance of asking questions is to ask students: 'What is the average of the numbers 5, 10, 350, 355?'

Most students will look suspicious but still say 180. The correct response is of course not an answer but another question: 'What are the numbers?' It is always unwise to analyse numbers without knowing what they are. In this case, on being told that the numbers are phase angles (so that $350°$ corresponds to $-10°$), it becomes clear that the sensible way to average the data is as 'zero degrees'.

It is arguable that the overall 'success' of many statisticians is largely determined by their effectiveness as statistical consultants. Unfortunately, it is difficult to teach effective consulting except by providing experience 'on the job'. However an increasing number of colleges now provide practical courses, variously labelled as 'Applied Statistics' or 'Statistical Consulting', which involve tackling substantial projects. These can give much useful guidance and training. Many possible projects and exercises are given in Part II of this book and some general instructions and advice are given in the Introduction thereto. The key point is that the student learns to tackle a **problem** rather than just being able to perform particular statistical techniques. For example, one possibility is to set questions which are deliberately incomplete or misleading (e.g. see Exercises B.3 and F.1) so as to emphasize the importance of clarifying objectives. (It is also instructive to point out illustrative statistical errors in the literature, so that students realize that a questioning attitude is judicious.) Another useful training scheme is to involve postgraduate students in genuine consulting sessions where they will quickly learn how to interact more effectively with people in other professions.

In some studies, particularly sample surveys and clinical trials, the statistician should be prepared for the **ethical problems** which may arise. For example, when is it ethical to prescribe a new treatment, or continue to prescribe an older treatment when a newer one is suspected of being better? In sample surveys, how does one decide what personal questions are reasonable to ask and under what circumstances? And to whom do the results of a survey 'belong'? A general code of ethics has been published by the International Statistical Institute (1986). This sets out the statistician's obligations to society, to funders and employers, to colleagues and to the human 'subjects'.

Finally, we mention **expert systems** (section 8.2) which attempt to mimic the interaction between a statistician and a user of statistics by means of an 'intelligent' computer package. One aim is to obviate the need for a statistical consultant in many situations. While some progress along these lines is clearly desirable so as to avoid the misuse of packages, it seems doubtful whether expert systems ever will (or should) take over the statistical consultant's role completely.

Further reading

Further advice on consulting is given by Sprent (1970), Rustagi and Wolfe (1982) and Joiner (1982). Hand and Everitt (1987) have edited a collection of examples showing the statistical consultant in action of which Tony Greenfield's reminiscences are particularly entertaining.

11
Communication — 2: Effective report writing

A good statistician must be able to communicate his work effectively, both verbally and by means of a written report. There is little point in 'doing a good job' unless the results can be understood by the intended recipients (see Exercise G.2).

This chapter concentrates on giving general guidelines for writing a clear, self-contained report which is the normal method of communicating results. This can be a difficult and sometimes tedious job. However, written documentation of your work is vital and should be done before memory fades. The outcome of a project is often judged by what is written rather than by what was actually done!

An alternative method of communicating results is by means of an oral presentation. This allows discussion and feedback and so is especially suitable for interim presentations. Good visual aids are essential and some brief guidelines are given in section 11.4.

The three main stages in writing a report may be described as **preparation**, **writing** and **revision**, and they are considered in turn.

11.1 Preparation

Before you start writing a report, you should collect together all the facts and ideas about the given topic which you want to include in the report. Sketch a brief outline of all the different points which need to be included, and plan the structure of the report. This involves getting the material into the right order and dividing it into sections (and possibly subsections) which should be numbered consecutively. Give each section a suitable heading; common titles include 'Introduction', 'Description of the experiment', 'Discussion of results' and 'Conclusions'.

11.2 Writing the report

Statisticians have widely different abilities to express themselves in writing. Yet, by following simple general principles and acquiring good technique, the reader should be able to produce a report of a reasonable standard. Before you start to write, consider carefully who is going to read the report, what their level of knowledge is likely to be, and what action, if any, you want the report to precipitate. The following general guidelines should be helpful:

1. Use simple, clear English. In particular, short words should be preferred to long words with the same meaning. Try to avoid sentences which are longer than about 20–25 words, by splitting long sentences in two if necessary.
2. If you cannot think of exactly the right word, a reference book of synonyms or a thesaurus may be helpful.
3. Use a dictionary to check spelling.
4. Add sufficient punctuation, particularly commas, to make the structure of each sentence clear.
5. Important words or phrases may be <u>underlined</u> or written in **bold**, in *italics* or in CAPITALS to make them stand out.
6. The hardest step is often to 'get started' at all. The first word is usually the most difficult to write. A thorough preparation (section 11.1) is very helpful as there may be little distinction between jotting down preliminary ideas and the first draft. Try writing the first draft as if telling a friend about your work in your own words. You can then polish the style later. It is much easier to revise a draft (however bad) than to write the draft in the first place.
7. It is often easier to write the middle sections of the report first. The introduction and conclusions can then be written later.
8. The introduction should provide a broad, general view of the topic. It should include a clear statement of objectives and indicate how far they have been carried out. Some background information should be given but avoid details which belong later in the report.
9. The conclusions should summarize the main findings and perhaps recommend appropriate action.
10. A brief summary or abstract at the beginning of the report is often useful.
11. The summary, introduction and conclusions need to stand on their own and be particularly clear, as some readers may only look at these sections.
12. Graphs and tables form an important part of many reports. They require careful preparation but sadly are often poorly produced. Section 6.5 gives general advice on presentation. In brief, they should have a clear, self-explanatory title, so that they are understandable when viewed alone, and they should be numbered so that they can be referred to in the text. The units of measurement should be stated. The axes of graphs should be labelled. Tables should not contain too many (or too few) significant digits. Do not include undigested computer output, as the graphs and tables therein may be quite unsuitable for presentation purposes. Computer tables are often poorly formatted and labelled, while computer graphs can look hideous with poorly designed axes and unclear labelling. It is sometimes easier to revise computer output with the help of correcting fluid and add

labels/comments by hand, but it is obviously more satisfactory to use good software which enables the user to control the output.

13. Appendices are useful for detailed material which would break up the flow of the main argument if included in the main text. This includes detailed mathematics and large tables of computer output (if they are really necessary at all). The latter should be summarized in the main text.
14. The technical details in the main sections should be clear, concise and mathematically sound. Define any notation which is introduced. Give sufficient background theory but do not try to write a book!
15. A bibliography is often desirable, and references to books and papers should be given in full so that they can easily be found. A reference should include the author's name(s) and initials, date of publication, title, publisher (for a book) or name of journal, volume and page numbers (for a paper), as illustrated in the references at the end of this book. Get into the habit of recording the full details of a reference at the time you use it.
16. The report should be dated, given a title, and say who has written it.

11.3 Revision

When you have finished the first draft of the report, you will find that substantial revision is usually necessary. First, ask yourself if the arrangement of sections (the structure) is satisfactory. Second, examine the text in detail. Does it read easily and smoothly? Is it clear? Is there unnecessary repetition? Have any important points been omitted? Are the graphs and tables clear? The report can be improved substantially by generally 'tightening' the style. You need to be as brief as possible while still including all important details. You should also double-check any numbers given in the text, particularly if they have been copied several times.

Do not underestimate the time required to revise a report, which often exceeds the time taken to write the first draft.

When you think you have finished, put the report aside for at least 24 hours, and then read it through in one sitting. Try to imagine yourself as a reader who is seeing it for the first time. You may even find a sentence which you yourself cannot understand! It is amazing how many obvious errors are left in reports because they have not been properly checked. You will be judged on what you have written and not on what you meant to write!

Before getting a report typed, try and get someone else (your supervisor?) to read it. Do not be surprised, angry or discouraged if you are advised to make extensive changes, as this is the fate of most draft reports (including the first version of this book!). Most secretaries know little statistics, and need guidance with formulae containing unusual symbols, such as Greek letters. Obviously fewer typing errors will occur if you write legibly. After typing, check the typescript carefully and in particular check formulae symbol by symbol.

Remember that if a mistake gets through, then it is your fault and not the secretary's!

11.4 Oral presentations

Many people are required to make oral presentation during their career. This can be disturbing, threatening or even terrifying if you are not used to it. Careful preparation and practice is the key to success.

Good visual aids are vital and are expected by the sophisticated audiences of today. Usually they consist of transparencies for an overhead projector (OHP). Write clearly in large letters or get them typed in a larger size than normal. Nowadays, word processors enable good-quality printed transparencies to be produced very easily. Make sure they can be read at the back of the room in which you will be making your presentation. Note that graphs (and other material) can be photocopied directly onto special acetate paper. A common mistake is to put too much material on one slide so that it becomes visually off-putting or even illegible. This should be avoided.

Slides are for the more advanced lecturer as they are harder to produce and in my experience trickier to handle, as a slide projector is more likely to jam or go wrong when changing slides.

Have a timed practice, including the use of your visual aids, preferably with an audience of at least one who is prepared to comment critically on your performance. If you have, say, 20 minutes to speak, you should make sure you speak for your allotted time and nor for 10 minutes or 40.

Well before you start, check that you know how to operate any visual aid machines and that they are focused. There is nothing more annoying for an audience than seeing a speaker introduced who then spends the first five minutes of the talk fiddling with the OHP. Another obvious good piece of advice (which is often ignored!) is to make sure that you do not stand in front of the screen and prevent the audience seeing your wonderful visual presentation.

Further reading

There are many specialized books on report writing. It is worth looking at the classic text by Sir Ernest Gowers (1986). Other useful references include Wainwright (1984) and Ehrenberg (1982, Chapter 18).

12
Numeracy

As well as learning statistical techniques, the aspiring statistician needs to develop a sense of **numeracy**. This is rather hard to define and cultivate, but involves general numerical 'common sense', some knowledge of probability and statistics, sound judgement and the ability to 'make sense' of a set of numbers.

This chapter introduces two other aspects of numeracy, namely the need to maintain a healthy scepticism about other people's statistical analyses and conclusions, and the importance of being able to deal with the sort of misconceptions which typically arise in the minds of the 'general public'.

12.1 The need for scepticism

> Public agencies are very keen on amassing statistics – they collect them, raise them to the nth power, take the cube root, and prepare wonderful diagrams. But what you must remember is that every one of the numbers comes in the first instance from the village watchman, who just puts down what he damn well pleases. *Sir Josiah Stamp*

The sensible statistician should be wary of accepting, without question, any statistical results, conclusions or formulae that they might see in print. For example, it is unwise to believe all official statistics, such as government assertions that the probability of a nuclear meltdown is only one in 10 000 years (remember Chernobyl!). As another example, the number of unemployed is a politically explosive weapon and it might be thought that an index like this should be above suspicion. In fact it is quite difficult to define what is meant by an unemployed person. Should this term apply to someone working part-time, to people who are unfit for work, to married women who do not want to go out to work, and so on? In the United Kingdom, the method of calculating the number of unemployed has been changed several times in recent years. In each case the change has miraculously reduced the number of unemployed! Example 4.6 is another example of how official statistics can be made up or just be plain wrong. Governments may be trying to collect too much information of too complicated a format so that mistakes become inevitable. How, for example, should a social worker fill in a form asking for the 'main purpose of a visit' made to a family of two adults and three young children sleeping in two beds and living in one room where state benefit has failed to arrive, the mother is unwell and one child has a worrying bruise?

A healthy scepticism means that the statistician should also look out for mistakes in books and journals. Misprints and errors do get published. For example, on reading that a group of 23 medical patients have an average age of 40.27 years with range 24.62 years and standard deviation 2.38 years, the reader should realize straight away that the last two statistics are almost certainly incompatible – despite the apparent two-decimal-place accuracy (which is excessive anyway!). Perhaps the 'standard deviation' is meant to be the standard error of the mean? Unfortunately, it is easy to quote many other alarming mistakes from published work, including incorrect formulae as well as statistics (see Example 7 of Chatfield, 1991). Some such mistakes are just typographical errors and most books (including this one!) will inevitably contain some examples of this type of error, not all of which need be the fault of the author. Nevertheless, some textbooks undoubtedly contain more misprints than can reasonably be expected due to typesetting problems. For example, I recently saw a textbook which gave the linear regression model relating a response variable, y, to an 'independent' variable, x, as

$$y = b_0 + b_1 + \varepsilon$$

It repeated the omission of the x-variable in later equations and then presented incorrect formulae for the estimates of b_0 and b_1. Readers of that book will have difficulty!

The only answer to the problem of misprints and errors is eternal vigilance, and the following checks are advisable when scrutinizing formulae in published work:

- Check the **dimensions** of a formula (Edwards and Hamson, 1989, p. 60). Thus if observations are lengths (L), then the mean and standard deviation should also be lengths (L), but variance should be a squared length (L^2).
- Check the result given by the formula in a (simple) special case where you know the correct answer beforehand.
- Check a formula by referring to a standard text or to an experienced statistician.
- Check the **limiting behaviour** of a formula where possible. For example, you may know what the result should be as one variable approaches zero or infinity.

As well as mistakes in formulae and in numerical values, it is all too easy to find examples where an **incorrect form of analysis** has been adopted. The widespread availability of computer software means that the non-specialist (or even the specialist!) can commit a much wider range of offences. Fortunately, we can sometimes learn as much from mistakes (our own and other people's) as we do from taking lecture courses and reading textbooks. A list of some common types of blunder which could be made during a statistical analysis is given below:

1. The silly statistic: if, for example, a sample mean is calculated which is outside the range of the given data, then there is obviously an arithmetical error. More subtle is the misleading use of a sample mean to describe bimodal or censored data, and the inappropriate use of the standard deviation to describe skewed data.
2. The silly graph: a graph with no title or unlabelled axes.
3. The paired comparison test: with paired difference data, one of the commonest errors in the literature is to carry out a two-sample test on means rather than the appropriate paired difference test. Then the treatment effect is likely to be swamped by the differences between pairs.
4. The silly regression: fitting a straight line to data which are clearly non-linear.
5. Another silly regression: with some computer packages, regression through the origin gives a higher coefficient of determination (R^2) than fitting a model with a constant term, even though the latter contains one more parameter. The user may therefore be tempted to use regression through the origin all the time! This problem arises when sums of squares are calculated about the origin rather than about the mean.
6. The silly χ^2-test: a frequent howler is to carry out the χ^2 goodness-of-fit test on a table of percentages, or proportions, rather than on the table of count frequencies.
7. The silly P-value: significance tests are often carried out on silly null hypotheses (which are obviously untrue anyway) using messy data which do not in any case satisfy the required conditions for the test to be valid.
8. The silly multiple regression: where the number of explanatory variables is close to (or even bigger than) the number of observations.
9. The silly experimental design or silly sample survey: some painful examples were given in Chapter 4. Finding out how data were collected is just as important as analysing them.

This list could easily be extended. The reader must be wary of statistical analyses carried out by other people and not assume they are correct. It is encouraging that some journals are attempting to improve the presentation of statistical results. In particular, the *British Medical Journal* has laid down guidelines (Altman *et al.*, 1983) on what to include when writing papers. These guidelines could usefully be imitated by other journals.

12.2 Dealing with popular misconceptions

Statistical reasoning can affect nearly all aspects of our lives. Thus, apart from dealing with 'professional' statistics, it is important for statisticians to be sufficiently numerate to cope with the flood of statistics and pseudo-statistics from the media and the sort of misconceptions embraced by the general public. Given so many statistics of such variable quality, it is unfortunate that many people either believe everything they hear or (perhaps even worse) come to believe in nothing statistical.

In order to understand the sort of misconceptions which can and do arise, and learn how to spot 'daft' statistics, it is a good idea to read the light-hearted books by Huff (1959; 1973). Some other good books of a similar style include Hooke (1983) and Hollander and Proschan (1984). A few brief examples must suffice here.

One popular misconception is that anyone who handles numbers must be a statistician. This is like saying that anyone who handles money must be a banker. In fact most statistical 'lies' are produced by non-statisticians.

Another popular misconception is that random variability is somehow abnormal. In fact one of the first lessions in statistics is that random variability is perfectly normal and needs to be measured and understood so as to allow the estimation of other more interesting effects. The general public needs to learn that the average can be overworked and should usually be supplemented by some measure of spread. For example, people *are* different! It is much better for a doctor to tell a patient that recovery will take between, say, three and seven days, than to say that it will take five days on average. Anyone who takes longer than five days will get worried!

A third misconception is that any data are good data. In fact the statistician knows that data collection is just as important as data analysis, and that stopping the first 50 people to emerge from the local supermarket on a Saturday morning will not give a random sample of consumers.

There are many popular misconceptions regarding probability. For example, television commentators often use the word 'certain' when they mean 'probable' (e.g. the footballer was certain to score until he missed!). The law of averages is also widely misquoted. To take just one example, a newspaper reported a team captain as having lost the toss four times in a row before important matches and therefore 'by the law of averages' was more likely to win the next toss!

The final misconception we mention here is that all official statistics are 'true'. This has already been covered in section 12.1.

We close with three examples of the sort of potentially misleading or plain daft statistics foisted on the general public.

1. 'More accidents happen in the home than anywhere else.' This is an example of 'so-what' statistics. Most people spend most of their time at home and so the result is hardly surprising. We cannot all move away from home to avoid accidents!
2. 'Sales are up 10%.' This is an example of the **unmentioned base**. If sales were poor before, then they are not much better now.

3. '67% of consumers prefer brand X.' This is an example of the **unmentioned sample size**. If a sample of size three has been taken, the result is hardly convincing.

No doubt the reader can easily add to this list (e.g. Exercise A.3). The main lesson is to be vigilant at all times.

12.3 Tailpiece

Having emphasized the importance of numeracy, I close paradoxically by reminding the reader that many of the most important things in life are difficult or impossible to measure; for example, beauty, joy, love, peace, and so on. Thus cost-benefit analysis, which was very fashionable a few years ago, has been described as 'a procedure by which the priceless is given a price ... an elaborate method of moving from pre-conceived notions to foregone conclusions' (Schumacher, 1974). The good statistician must be prepared to use his subjective judgement where necessary to modify the results of a formal statistical analysis.

Summary
How to be an effective statistician

We conclude Part I with a brief summary of the qualities needed by an effective statistician and of the general principles involved in tackling statistical problems.

The attributes needed by a statistician are a mixture of technical skills and beneficial personal qualities. A 'good' statistician needs to be well trained in both the theory and practice of statistics and to keep up with the statistical literature. The theory should include an understanding of probability models and inference, and some mathematical ability is useful. The practice should include plenty of experience with real data in a variety of disciplines so as to develop a 'feel' for the realities of statistical life. In particular, the statistician should appreciate the importance of all the different stages of a statistical investigation, and understand the general principles involved in tackling statistical problems in a sensible way. The latter may be summarized as follows:

1. When a study is proposed, **formulate the problem** in a statistical framework. **Clarify the objectives** carefully. **Ask questions** to get sufficient background information on the particular field of application. Search the literature if necessary.
2. Scrutinize the method of **data collection** carefully (especially if it has not been designed by a statistician). It is important to realize that real data are often far from perfect.
3. Look at the data. An **initial data analysis (IDA)** is vital for both data description and model formulation. Be aware that errors are inevitable when processing a large-scale set of data, and steps must be taken to deal with them.
4. Choose and implement an appropriate method of analysis at an appropriate level of sophistication. Rather than asking 'What technique shall I use here?' it may be better to ask **'How can I summarize these data and understand them?'**. Rather than thinking of the analysis as just 'fitting a model', there may be several cycles of model formulation, estimation and checking. If an effect is 'clear', then the exact choice of analysis procedure may not be crucial. A simple approach is often to be preferred to a complicated approach, as

the former is easier for a 'client' to understand and is less likely to lead to serious blunders.

5. Be ready to adapt quickly to new problems and be prepared to make *ad hoc* modifications to existing procedures. Be prepared, if necessary, to extend existing statistical methodology. Being prepared to use **lateral thinking**, as the 'best' solution to a problem may involve looking at it in a way which is not immediately apparent, or even answering a different question to the one that was originally posed.

6. If the problem is too difficult for you to solve, do not be afraid to consult a colleague or an appropriate expert.

7. Write a convincing report, which should be carefully planned, clearly written and thoroughly checked.

8. Do not just react to situations, but look actively for ways to improve things.

In addition to technical knowledge, the ability to make technical judgements, and plenty of practical experience with real data, the ideal statistician would also have the following more personal qualities:

(a) be an effective **problem solver**;

(b) be **thorough** and yet report on time;

(c) be **open-minded** and display a **healthy scepticism**;

(d) be able to **collaborate** with other people, and be able to **communicate** both orally and in writing;

(e) be versatile, adaptable, resourceful, self-reliant and have sound common sense (a tall order!);

(f) be able to use a **computer** and a **library** effectively;

(g) be **numerate**, particularly in being able to 'make sense' of a set of messy data, yet also understand that some important facets of life cannot be expressed numerically and hence understand what statistics can and cannot do.

The paragon of virtue depicted here is very rare, but should still provide us with a target to aim at.

PART TWO
Exercises

This part of the book presents a varied collection of exercises, ranging widely from fairly small-scale problems to substantial projects involving real-life, complex, large-scale data sets. There are also some exercises which involve collecting data, using a library and report writing.

The exercises are designed to illustrate the **general principles** described in Part One of the book for tackling real-life statistical problems, rather than the use of **techniques** as in most other textbooks. With a few exceptions, the exercises are generally posed in the form of a real-life problem rather than as a statistical or mathematical exercise. While some of the analyses turn out to have a 'standard' form, the method is usually not specified in the question and so part of the problem is to translate the given information into a statistical formulation. In particular, the exercises illustrate the importance of:

- clarifying objectives and getting background information;
- finding out how the data were collected;
- carrying out an initial data analysis (IDA);
- formulating an appropriate model;
- improvising so as to cope with non-standard data and unfamiliar problems;
- trying more than one approach where necessary;
- presenting the results clearly;
- avoiding trouble.

I hope the exercises give the reader experience in thinking about, and then tackling, real practical problems. This should help develop the practical skills needed by a statistician. I do not claim that the exercises are a 'random sample' of real statistical problems (if such a thing were possible), but my experience does suggest that they are more representative of 'real life' than those in most other statistics books.

Each exercise is designed to make at least one important point. Except where otherwise indicated, the data are real and peculiarities have generally been left in. Such peculiarities help to illustrate the importance of finding out exactly how data have been collected and emphasize the need for early consultation with a statistician so as to get a better-designed study and hence 'better' data. Some simplification has occasionally been made in order to get exercises of a reasonable length, which are suitable for student use, but this has been kept to a minimum. The dangers of oversimplification, and of giving insufficient background information in illustrative examples, are noted by Preece (1986). I also note that some non-standard exercises (e.g. Exercises A.1–A.4, C.1 and C.2) may appear elementary, but are designed to be entertaining and yet make the reader think and ask questions.

Most published examples in the literature are idealized and sanitized to avoid mentioning difficulties, mistakes and blind alleys, and do not make clear the iterative nature of much real statistical analysis. This is a pity. We can learn from our mistakes as well as from our successes. Indeed Tony Greenfield's article in Hand and Everitt (1987) says that 'many true stories about consultation are tales of woe, but these are where the lessons are to be learnt'. Thus I have not attempted to 'clean up' artificially the data, the problems or the 'solutions'. I have nevertheless generally avoided setting questions which are deliberately incomplete or wrong (but see Exercise B.3), although it could be argued that such exercises are needed to illustrate real life. Teachers may wish to add some such exercise of their own. Students certainly learn fast that they need to ask questions when an incomplete exercise is given!

Tackling the exercises

The exercises have been divided into sections with self-explanatory titles and use a question-and-answer format. Within each section, the solutions (or helpful comments) are presented in order at the end of the section so that the reader cannot see both problem and solution at the same time. This is quite deliberate. You should read the problem, think about it, and hopefully tackle it before reading the solution. **Do not cheat!**

You are advised to examine carefully each data set before embarking on a sophisticated analysis. This should always be done anyway. In some cases the outcome of the IDA is so clear that a definitive analysis is unnecessary (or even undesirable). In other cases, the outcome of the IDA may not be clear or the data structure may be so complicated that a more sophisticated analysis is necessary. Even so, the IDA should be helpful in choosing the method of analysis.

The solutions

These are not meant to be definitive but are often left open-ended to allow the reader to try out different ideas. Some readers may query some of my 'comments' and alternative solutions can doubtless be found. (We all like to think we can analyse data a little better than everyone else.) I have perhaps overemphasized IDA and underemphasized the use of standard statistical techniques, but I regard this as a desirable reaction to the current overuse of too complicated methods. Knowing lots of theory may make it easier for a statistician to go off in the wrong direction and I certainly hope the exercises demonstrate the vast potential for simple ideas and techniques.

Links with other practical work

The exercises have been divided into sections with self-explanatory titles and use a question-and-answer format. Within each section, the solutions (or helpful courses, most exercises are what I call **technique-oriented** in that they drill students on the use of a particular statistical method. The student is typically told exactly what to do with a statement of the form: 'Here are some data. Apply technique X.' Such exercises are an essential step in learning techniques, but do not prepare students for real life and the likelihood that data will not arrive on the statistician's desk in a neat and tidy form with precise instructions as to the appropriate form of analysis. Thus the objectives of the exercises in this book are quite different. They are primarily what I call **problem-oriented** exercises, in that the reader is given a problem to solve and has to decide for him or herself how to tackle it. These exercises are more likely to be of the basic form: 'Here are some data. Analyse them.' Sometimes it is not even clear what the aim of the analysis is, and part of the problem-solving strategy will involve asking questions to clarify objectives and getting appropriate background information. This is rather hard to simulate in a book setting but some pointers are given. However, the exercises concentrate on the **choice** of an appropriate form of analysis for a given problem. This can be difficult and so many of the problems have dual titles describing both the subject matter and a hint about the method of analysis. Even so, I hope the student will learn to ask questions such as 'What background knowledge do I need here?', and 'How can I summarize these data and understand them?', rather than just 'What technique shall I use here?'.

Alternative types of practical work are reviewed by Anderson and Loynes (1987, Chapter 3). They include **experiments** where students collect their own data (see Chapter H in this book), **case studies** where students are taken step-by-step through a real problem, and **projects** (e.g. Kanji, 1979) where students undertake a fairly substantial piece of work. Projects make students work by themselves, find out how to use a library effectively, carry through an investigation from start to finish, and write up a clear report. I would argue that a series of small-scale projects, as in Part Two here, will give a student a wider variety of experience than a single large-scale project, though there is certainly a place for the latter, and I would recommend that at least one of the larger exercises in this book should be tackled individually, analysed thoroughly, and written up properly. Further exercises of a related kind may be found, for example, in Cox and Snell (1981) and Anderson and Loynes (1987). Also note that over 500 small data sets from a wide variety of backgrounds and data types are given by Hand *et al.* (1994), and they could easily be used to construct additional exercises.

Some remarks on the teaching of applied statistics

Most education concentrates on **amassing knowledge**, which includes learning the details of statistical techniques. However, there is increased interest in **experiential learning** and getting students to **think** and **solve problems**. It is arguably much more important to learn to think than just to be stuffed full of facts. However, some knowledge is obviously a prerequisite to being able to think constructively. Thus good statistics teaching will present a sensible blend of techniques and strategy (as well as a sensible blend of theory and practice). For example, the increasing use of computer software to carry out statistical techniques means that the user no longer needs to know all the algebraic details of a method (though he/she must know the underlying model and assumptions on which it is based) but rather needs help to be able to select and use appropriate software and then be able to interpret the output.

I suggest that an introductory statistics course (whether for specialists or for non-specialists) should begin by introducing descriptive statistics, as this provides an intuitive basis for inference. However, after learning inference, I suggest that the student should 'go back' to look at IDA again and learn to differentiate situations where IDA indicates that no formal analysis is necessary or where IDA helps to formulate a sensible model and hence indicate an appropriate analysis. This can only be done properly when the student has sufficient prior knowledge about modelling and inference. When the time is right, I also suggest that the student is explicitly taught some general strategies for problem solving as in Part One of this book. Indeed, given that the 'success' of many statisticians is determined by their effectiveness as statistical consultants, I believe that a course on statistical consulting, statistics problem solving, applied statistics or some such similar title should be part of the training of every statistician. This book is designed to be used by advanced undergraduates or postgraduates as a text for such a course. Most of the exercises in Part Two have been tried on final-year undergraduates at the University of Bath as part of a course examined by continuous assessment rather than by a written exam. This course also includes some lectures on the general principles of tacking real-life statistical problems, as covered in Part One of the book.

A

Descriptive statistics

Descriptive statistics is an important part of the initial examination of data (IDA). It consists primarily of calculating summary statistics and constructing appropriate graphs and tables. It may well be regarded by the reader as the most familiar and easiest part of statistics. Thus the exercises in this section are intended to demonstrate that descriptive statistics is not always as easy as might be expected, particularly when data exhibit skewness and/or outliers. It is not always clear what summary statistics are worth calculating, and graphs and tables are often poorly presented.

Further relevant exercises include Exercises B.1, B.2 and B.3 which demonstrate the power of a good graph, and Exercises B.4 and B.6 which give further hints on presenting a clear table. Exercises D.1 and D.2 are also relevant, while the importance of the time plot in time-series analysis is illustrated in Chapter F. In fact nearly all the exercises involve some descriptive statistics as part of IDA.

Exercise A.1 Descriptive statistics – I

The 'simplest' type of statistics problem is to summarize a set of univariate data. Summarize the following sets of data in whatever way you think is appropriate.

(a) The marks (out of 100 and ordered by size) of 20 students in a mathematics exam:

$$30, 35, 37, 40, 40, 49, 51, 54, 54, 55$$
$$57, 58, 60, 60, 62, 62, 65, 67, 74, 89$$

(b) The number of days work missed by 20 workers in one year (ordered by size):

$$0, 0, 0, 0, 0, 0, 0, 1, 1, 1$$
$$2, 2, 3, 3, 4, 5, 5, 5, 8, 45$$

(c) The number of issues of a particular monthly magazine read by 20 people in a year:

$$0, 1, 11, 0, 0, 0, 2, 12, 0, 0$$
$$12, 1, 0, 0, 0, 0, 12, 0, 11, 0$$

(d) The heights (in metres) of 20 women who are being investigated for a certain medical condition:

$$1.52, \ 1.60, \ 1.57, \ 1.52, \ 1.60, \ 1.75, \ 1.73, \ 1.63, \ 1.55, \ 1.63$$
$$1.65, \ 1.55, \ 1.65, \ 1.60, \ 1.68, \ 2.50, \ 1.52, \ 1.65, \ 1.60, \ 1.65$$

Exercise A.2 Descriptive statistics – II

The following data are the failure times in hours of 45 transmissions from caterpillar tractors belonging to a particular American company:

4381	3953	2603	2320	1161	3286	6914	4007	3168
2376	7498	3923	9460	4525	2168	1288	5085	2217
6922	218	1309	1875	1023	1697	1038	3699	6142
4732	3330	4159	2537	3814	2157	7683	5539	4839
6052	2420	5556	309	1295	3266	6679	1711	5931

Display a sensible stem-and-leaf plot of the data and from it calculate the median and interquartile range. Without calculating the mean, say whether it is greater than or smaller than the median.

Construct the boxplot of the data. Do you think it is preferable to display these data using (a) a histogram, (b) a stem-and-leaf plot or (c) a boxplot?

Describe the shape of the distribution of failure times and indicate any observations which you think may be outliers. Find a transformation such that on the transformed scale the data have an approximately symmetric distribution, and comment again on possible outliers.

Exercise A.3 Interpreting 'official' statistics

One class of 'descriptive statistic' is formed by the wide variety of national and international statistics. They are apparently easy to interpret, or are they?

(a) Discuss the following statements:
   (i) In recent years, Sweden has had one of the highest recorded suicide rates. This indicates problems in the Swedish way of life.
   (ii) UK official statistics show that women giving birth at home are more at risk than women giving birth in hospital. This indicates that all babies should be delivered in hospital.
   (iii) On comparing death rates from tuberculosis in different states of the USA, it is found that Arizona has the worst record in recent years. This indicates that Arizona is an unhealthy place to live.
(b) The railways of Great Britain have always set high standards of safety. A spate of serious accidents in 1984 suggested that safety standards might

Table A.1 Railway accidents on British Rail, 1970–83

Year	No. of train accidents	Collisions Between passenger trains	Collisions Between passenger and freight trains	Derailments Passenger	Derailments Freight	No. of train miles (millions)
1970	1493	3	7	20	331	281
1971	1330	6	8	17	235	276
1972	1297	4	11	24	241	268
1973	1274	7	12	15	235	269
1974	1334	6	6	31	207	281
1975	1310	2	8	30	185	271
1976	1122	2	11	33	152	265
1977	1056	4	11	18	158	264
1978	1044	1	6	21	152	267
1979	1035	7	10	25	150	265
1980	930	3	9	21	107	267
1981	1014	5	12	25	109	260
1982	879	6	9	23	106	231
1983	1069	1	16	25	107	249

Source: Chief Inspecting Officer of Railways Annual Reports, Department of Transport. Figures given may be slightly influenced by changes in statistical treatment, but they have been allowed for where possible and are thought not to affect the conclusions significantly.

have deteriorated. Is there any evidence from the data given in Table A.1 that standards are declining?

Exercise A.4 Lies, damned lies ...

Capital punishment for murder in the United Kingdom was provisionally abolished in 1965. Permanent abolition followed in 1969, but only after a lively debate. During this debate a national newspaper published the graph shown

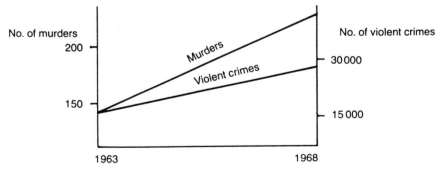

Figure A.6 A graph from a newspaper.

in Fig. A.6 to support the case for retaining the death penalty. Comment on the graph.

Notes on Exercise A.1

Data sets may be summarized graphically, numerically, and/or verbally. The choice of appropriate summary statistics depends in part on the shape of the underlying distribution. Assessing shape is therefore an important first step in data description. This can be achieved by drawing a bar chart (for a discrete variable), a histogram (for a continuous variable), or a stem-and-leaf plot.

(a) The exam marks data are the only straightforward data set. The histogram in Fig. A.1 reveals a reasonably symmetric bell-shaped distribution. As the data are approximately normally distributed, suitable summary statistics are the sample mean, $\bar{x} = 55$ marks, as a measure of location, and the sample standard deviation, $s = 14$ marks, as a measure of spread. The histogram, mean and standard deviation together provide a satisfactory summary of the data. Do not forget to give the unit of measurement (marks in this example) when quoting $\bar{x}$ and s.

Even in this simple case, there are pitfalls. Did you give too many significant figures in the summary statistics? When the data are recorded as integers there is no point in giving the summary statistics to more than one decimal place. As to the graph, you may have chosen a different width for the class intervals or given a different plot. For comparison, Fig. A.2 shows a stem-and-leaf plot with the smaller class interval width of five marks. This graph is like a histogram on its side with the first significant figure of each observation in the stem (on the left) and the second significant figure in the leaves (on the right). The digits within each class interval (or leaf) have been ordered by size. The reader can decide for himself which graph is the clearer.

(b) The bar chart in Fig. A.3 shows that the frequency distribution of 'days work missed' is severely skewed to the right. Note the break in the horizontal

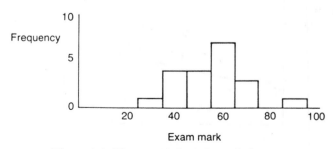

Figure A.1 Histogram of exam mark data.

```
3 | 0
3 | 5 7
4 | 0 0
4 | 9
5 | 1 4 4
5 | 5 7 8
6 | 0 0 2 2
6 | 5 7
7 | 4
7 |
8 |
8 | 9
```

THE STEM   THE LEAVES
UNITS = 10 marks   UNITS = 1 mark

Figure A.2 Stem-and-leaf plot of exam mark data.

axis between 8 and 45 days. The sample mean, $\bar{x} = 4.2$ days, is highly influenced by the largest observation, namely 45 days. The latter is an outlier, but there is no reason to think it is an error. By their nature, skewed distributions give outliers in the long 'tail'. The median (1.5 days) or mode (0 days) or the 5% trimmed mean (2.2 days) may be a better measure of location. The standard deviation is of little help as a descriptive measure of spread with such skewed data. The range is 45 days but this is also unhelpful when most of the observations lie between 0 and 5. The interquartile range, namely 4.7 days, may be preferred. Even so, summary statistics have limited value and the bar chart is probably the best way of summarizing the data.

(Note that the standard deviation may be of some use, not as a descriptive measure, but in finding a probability distribution to fit the data should this be desired. As the variance ($s^2 = 97.2$) is numerically so much larger than the mean, the Poisson distribution is (surprisingly?) not appropriate. Instead a distribution called the negative binomial may be appropriate.)

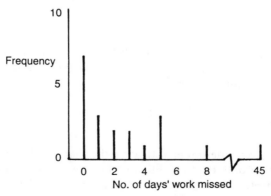

Figure A.3 Bar chart of absence data.

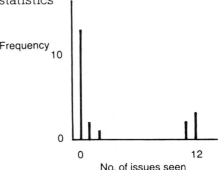

Figure A.4 Bar chart of magazine data.

(c) These data have not been ordered and are impossible to read 'by eye'. The first task is to construct the frequency distribution of 'number of issues read' which is plotted in Fig. A.4. There are two modes at 0 and 12. The bimodal U-shape is even more difficult to summarize than a skewed distribution. Most people do not read the magazine at all, but a substantial minority read nearly every issue. The sample mean and standard deviation are potentially very misleading. The proportion of 'regular' readers (5/20 or 25%) is a useful statistic, but it may be sensible to describe the data in words rather than with summary statistics.

(d) There was a misprint in these data which I have deliberately included in the problem. Did you spot it and deal with it? The observation 2.50 is not only an outlier, but almost certainly an error. It is probably meant to be 1.50 but you may prefer to omit it completely. As the remaining data are reasonably symmetric, they may be described by the sample mean and standard deviation. Another feature of the data is that, although they appear to be measured to two decimal places, inspection of the final digits suggests that some numbers, such as 1.65, keep recurring. A little detective work suggests that the observations have been measured to the nearest inch and converted to metres. Did you spot this?

Moral

Descriptive statistics is not always straightforward. In particular, the calculation of summary statistics depends on the shape of the distribution and on a sensible treatment of errors and outliers.

Notes on Exercise A.2

This exercise concentrates on investigating the shape of the underlying distribution of a given set of data. As highlighted by Exercise A.1, this is an important aspect of descriptive statistics, and a useful preliminary to the calculation of summary statistics.

```
0 | 2 3
1 | 0 0 1 3 3 3 7 7 9
2 | 2 2 2 3 4 4 5 6
3 | 2 3 3 3 7 8 9
4 | 0 0 2 4 5 7 8
5 | 1 5 6 9
6 | 0 1 7 9 9
7 | 5 7
8 |
9 | 5
```

THE STEM   THE LEAVES
UNITS = 1000 hours   UNITS = 100 hours

Figure A.5 Stem-and-leaf plot, with one-digit leaves, of transmission failure times.

The data in this example range from 218 hours to 9460 hours. A suitable width for each class interval in a stem-and-leaf plot is 1000 hours. This will give ten class intervals which is about 'right' for a sample size 45. The plot is easy to construct by hand, or using a package, and is shown in Fig. A.5. The leaves contain the second significant digit of each observation, ordered within each class interval. The length of each leaf is proportional to the frequency (cf. the corresponding histogram).

The median can easily be found from Fig. A.5 as the twenty-third observation, namely 3300 (or 3286 to be exact, from the raw data). The interquartile range is from 2200 (the twelfth observation) to 5100 (the thirty-fourth observation). As the distribution is skewed to the right, the mean exceeds the median. The boxplot has a 'box' from 2200 to 5100, with whiskers to the two extreme observations, namely 218 and 9460. The median is marked on the box at 3300. The boxplot loses much information and is generally unsuitable for a single sample (whereas a set of boxplots can be helpful for comparing several groups of observations, as in Exercise B.2).

The shape of the distribution is 'clearly' non-normal, but rather skewed to the right with a mode around 2000 hours. It is hard to assess outliers in the long 'tail' of a skewed distribution. Do you think the largest observation is an outlier? Although 2000 hours longer than the second highest observation, the long tail means there is no reason to think it is an error.

Using a computer, it is easy to try different transformations. In this case logarithms overtransform the data (making them negatively skewed), while square roots are 'about right' for giving approximate symmetry. In the histogram of the transformed data, the largest observation no longer looks 'outlying', whereas you may be surprised to find instead that the two smallest observations now look somewhat separated from the lower tail. In the absence of any external explanation why a square root transformation should be meaningful, the two lowest observations are still likely to be genuine, and I would say that there are no obvious outliers here. It is to be hoped that we now have a good understanding of the distribution of failure times, namely that it is skewed to the right with median 3300 hours and an interquartile range of 2900 hours

Notes on Exercise A.3

(a) These three statements demonstrate how easy it is to misinterpret official statistics. In (i) the word 'recorded' is important and this popular 'fact' (or myth?) about Sweden arises partly because the Swedes are more honest in recording suicide than many other nationalities. One or two countries actually have a zero suicide rate because suicide is not allowed as a legal cause of death! Any attempt to make inferences about the Swedish way of life would be very dangerous. In (ii) it is important to realize that there are two categories of home birth, the planned and unplanned. The former have low risk because of careful selection. The latter have high risk because they include premature and precipitate deliveries, etc., which will continue to occur at home whatever the official policy may be. Aside from ethical considerations, such as freedom of choice, statistics are needed on planned home deliveries in order to assess this question. As to (iii), Arizona is actually a healthy place for people who already have chest complaints to go to. Such people go to Arizona in sufficient numbers to boost the death rate artificially.

(b) These data are typical of many data sets in that a general sense of numeracy is more important for their interpretation than formal statistical training. As they are nationwide figures, there are no sampling problems, but there is still natural variability from year to year. The time series are too short to justify using formal time-series techniques. The wise analyst will begin by spending a few minutes just looking at the table. We see that the number of train accidents has reduced but then so have train miles, albeit by a smaller percentage. Collisions are few in number (thankfully!) and appear fairly random (are they approximately Poisson?). Freight derailments are substantially down, but passenger derailments, although much smaller in number, are not. If there is a decline in safety standards, it is not immediately obvious.

Is any further analysis indicated? There is no point in attempting any form of inference but it may be possible to clarify the table somewhat. Using the guidelines of section 6.5.2 of Part One, together with common sense, suggests (i) some rounding, (ii) calculating column averages (but not row averages!), (iii) reordering the columns, and (iv) grouping years to give fewer rows. Table A.1 (revised) is shown on p. 135. The general trends are now clearer. However, it is probably just as important to notice what data are not given and to ask questions to get further background information. For example, no information is given here on accident severity, although the total number of passenger fatalities is certainly of interest and must be recorded somewhere. No doubt further queries will occur to you.

Table A.1 (revised) Statistics related to accidents on British Rail, 1970–83

Year	No. of train accidents	No. of train miles (millions)	Collisions		Derailments	
			Between passenger trains	Between passenger and freight trains	Passenger	Freight
1970–74	1350	275	5.2	8.8	21	250
1975–79	1110	265	3.2	9.2	25	160
1980–83	970	250	3.7	11.5	23	107
Average	1160	265	4.0	9.7	23	177

Moral

When examining official statistics find out exactly what has been recorded and what has not been recorded. Take care in summarizing the data. The preparation of clear tables needs particular care.

Notes on Exercise A.4

The general principles of good graph design are given in section 6.5.3 of Part One. Figure A.6 is an appalling example of a graph deliberately designed to mislead the reader. The two scales have been chosen with false origins so that the line for murders appears to have a much steeper slope than that for violent crimes. In fact the percentage increase in murders is less than that for violent crimes. Other queries will no doubt occur to the reader. Why have the years 1963 and 1968 been chosen? Why draw a straight line between the two end years when intermediate figures are available? Criminal statistics can be difficult to interpret because of changes in the law and in the attitude of the courts. Murders are particularly hard to define. Many people who are indicted for murder are subsequently acquitted or found guilty of a lesser offence, such as manslaughter. Are these figures for murder convictions or what?

The total number of persons indicted for murder in England and Wales over the years 1957–76 are officially recorded as follows:

104, 104, 115, 124, 152, 135, 154, 158, 186, 247
223, 278, 249, 298, 309, 319, 361, 368, 394, 433

The overall increase in murder is, of course, very worrying, but it is difficult to detect evidence that the removal of capital punishment at about the middle of this period has had an appreciable effect.

Moral

As the media bombard the general public with statistics, tables and graphs, it is vital for numerate people to be on the lookout for the misuse of statistics and expose any 'horrors' which occur.

B
Exploring data

Given a new set of data, it is usually wise to begin by 'exploring' them. How many variables are there and of what type? How many observations are there? Are there treatments or groups to compare? What are the objectives? The initial examination of data (IDA), of which descriptive statistics is a part, is important not only to describe the data, but also to help formulate a sensible model. It is wise to explore data whether or not you think you know what analysis technique should be used. It is particularly important to check the assumptions made in carrying out a significance test, before actually doing it.

The examples in this section demonstrate how to cope with a varied selection of data sets, where the method of analysis may or may not be apparent. There are so many different types of data which can arise that it is quite unrealistic to suppose that you can always assess the appropriate method of analysis before closely examining the data.

Exercise B.1 Broad bean plants – a two-sample *t*-test?

The data given below show the (scaled) concentration of a certain chemical in 10 cut shoots of broad bean plants and in 10 rooted plants.

| Cut shoots: | 53 | 58 | 48 | 18 | 55 | 42 | 50 | 47 | 51 | 45 |
| Rooted plants: | 36 | 33 | 40 | 43 | 25 | 38 | 41 | 46 | 34 | 29 |

Summarize the data in whatever way you think is appropriate. From a visual inspection of the data, do you think there is a significant difference between the two sample means? Carry out a formal test of significance to see if the observed difference in sample means is significantly different from zero at the 1% level. Do any other questions occur to you?

Exercise B.2 Comparing teaching methods – an ANOVA?

In an experiment to compare different methods of teaching arithmetic, 45 students were divided randomly into five equal-sized groups. Two groups were taught by the currently used method (the control method), and the other three groups by one of three new methods. At the end of the experiment, all students took a standard test and the results (marks out of 30) are given in Table B.1 (taken from Wetherill, 1982, p. 263). What conclusions can be drawn about differences between teaching methods?

Table B.1 Test results for 45 students

Group A (control)	17	14	24	20	24	23	16	15	24
Group B (control)	21	23	13	19	13	19	20	21	16
Group C (praised)	28	30	29	24	27	30	28	28	23
Group D (reproved)	19	28	26	26	19	24	24	23	22
Group E (ignored)	21	14	13	19	15	15	10	18	20

Exercise B.3 Germination of seeds – an ANOVA?

Suppose that a biologist comes to you for help in analysing the results of an experiment on the effect of water concentration (or moisture content) on the germination of seeds. The moisture content was varied on a non-linear scale from 1 to 11. At each moisture level, eight identical boxes were sown with 100 seeds. Four of the boxes were covered to slow evaporation. The numbers of seeds germinating after two weeks were noted and are shown in Table B.2.

Table B.2 Numbers of seeds germinating

	Moisture content					
	1	3	5	7	9	11
Boxes uncovered	22	41	66	82	79	0
	25	46	72	73	68	0
	27	59	51	73	74	0
	23	38	78	84	70	0
Boxes covered	45	65	81	55	31	0
	41	80	73	51	36	0
	42	79	74	40	45	0
	43	77	76	62	*	0

* Denotes missing observation.

The biologist wants your help in analysing the data, and in particular wants to carry out an analysis of variance (ANOVA) in order to test whether the number of seeds germinating in a box is affected by the moisture content and/or by whether or not the box is covered. Analyse the data by whatever method you think is sensible. Briefly summarize your conclusions.

Exercise B.4 Hair and eye colour/family income and size – two-way tables

(a) The data in Table B.3 show the observed frequencies of different combinations of hair and eye colour for a group of 592 people (Snee, 1974). Summarize the data and comment on any association between hair and eye colour.

Table B.3 Observed frequencies of people with a particular combination of hair and eye colour

Eye colour	Hair colour			
	Black	Brunette	Red	Blond
Brown	68	119	26	7
Blue	20	84	17	94
Hazel	15	54	14	10
Green	5	29	14	16

(b) The data in Table B.4 show the observed frequencies of different combinations of yearly income and number of children for 25 263 Swedish families Cramér, 1946). Summarize the data and comment on any association between family income and family size.

Table B.4 Observed frequencies of Swedish families with a particular yearly income and family size

Number of children	Yearly income (units of 1000 kronor)				
	0–1	1–2	2–3	3+	Total
0	2 161	3 577	2 184	1 636	9 558
1	2 755	5 081	2 222	1 052	11 110
2	936	1 753	640	306	3 635
3	225	419	96	38	778
≥4	39	98	31	14	182
Total	6 116	10 928	5 173	3 046	25 263

Exercise B.5 Cancer in rats – truncated survival times

An experiment was carried out on rats to assess three drugs (see data set 3 of Cox and Snell, 1981, p. 170). Drug D is thought to promote cancer, drug X is thought to inhibit cancer, while P is thought to accelerate cancer. Eighty rats were divided at random into four groups of 20 rats and then treated as follows:

Group	Drugs received
I	D
II	D, X
III	D, P
IV	D, X, P

The survival time for each rat was noted and the results are given in Table B.8. A post mortem was carried out on each rat, either when the rat died, or at 192 days when the experiment was stopped. The letter N after a survival time means that the rat was found not to have cancer. Summarize the data and try to assess whether the three drugs really do have the effects as suggested.

Table B.8 Survival times in days for four groups of rats

Group I; D		Group II; DX		Group III; DP		Group IV; DXP	
18 N	106	2 N	192 N	37	51	18 N	127
57	108	2 N	192 N	38	51	19 N	134
63 N	133	2 N	192 N	42	55	40 N	148
67 N	159	2 N	192 N	43 N	57	56	186
69	166	5 N	192 N	43	59	64	192 N
73	171	55 N	192 N	43	62	78	192 N
80	188	78	192 N	43	66	106	192 N
87	192	78	192	43	69	106	192 N
87 N	192	96	192	48	86	106	192 N
94	192	152	192	49	177	127	192 N

Exercise B.6 Cancer deaths – two-way tables of proportions

The data shown in Table B.10 are taken from a cohort study into the effect of radiation on the mortality of survivors of the Hiroshima atom bomb (see data set 13 of Cox and Snell, 1981, p. 177). This exercise requires you to carry out an initial examination of the data and report any obvious effects regarding the incidence of death from leukaemia and death from 'all other cancers'. You are not expected to carry out a 'proper' inferential analysis even if you think this is desirable.

In practice, an analysis like this should be carried out in collaboration with appropriate medical experts, but here you are simply expected to use your

Table B.10 Number of deaths from leukaemia and from other cancers during the period 1950–59 for the given sample size alive in 1950

Age in 1950		Total	Radiation dose in rads					
			0	1–9	10–49	50–99	100–199	200+
5–14	Leukaemia	14	3	1	0	1	3	6
	All other cancers	2	1	0	0	0	0	1
	Alive 1950	15 286	6 675	4 084	2 998	700	423	406
15–24	Leukaemia	15	0	2	3	1	3	6
	ALL other cancers	13	6	4	2	0	0	1
	Alive 1950	17 109	7 099	4 716	2 668	835	898	893
25–34	Leukaemia	10	2	2	0	0	1	5
	ALL other cancers	27	9	9	4	2	1	2
	Alive 1950	10 424	4 425	2 646	1 828	573	459	493
35–44	Leukaemia	8	0	0	1	1	1	5
	ALL other cancers	114	55	30	17	2	2	8
	Alive 1950	11 571	5 122	2 806	2 205	594	430	414
45–54	Leukaemia	20	9	3	2	0	1	5
	ALL other cancers	328	127	81	73	21	11	15
	Alive 1950	12 472	5 499	3 004	2 392	664	496	417
55–64	Leukaemia	10	2	0	2	2	1	3
	ALL other cancers	371	187	80	57	22	17	8
	Alive 1950	8 012	3 578	2 011	1 494	434	283	212
65+	Leukaemia	3	1	1	0	0	0	1
	ALL other cancers	256	119	59	48	13	10	7
	Alive 1950	4 862	2 245	1 235	935	232	123	92

common sense. You should mention any queries which come to mind and state any questions which you would like to ask the medical experts, as well as presenting your analysis of the data.

Exercise B.7 Vaccinating lambs – a two-sample *t*-test?

The growth of lambs can be seriously affected by parasitic diseases, which may depend in part on the presence of worms in the animals' intestines. Various vaccines have been proposed to reduce worm infestation. An experiment was carried out to investigate these vaccines, full details of which are given by Dineen, Gregg and Lascelles (1978). For financial reasons, each vaccine was investigated by a fairly small experiment in which unvaccinated lambs acted as controls and vaccinated lambs formed the treatment group. Each lamb was injected with worms and then samples were taken some weeks later. For one

Table B.12 Worms present in samples taken from vaccinated and unvaccinated lambs

	Sample size	Numbers of worms ($\times 10^3$)							
Control group	4	22,	21.5,	30,	23				
Treatment group	8	21.5,	0.75,	3.8,	29,	2,	27,	11,	23.5

vaccine, the data were as shown in Table B.12. Is there any evidence that vaccination has reduced the number of worms present?

Exercise B.8 Ankylosing spondylitis – paired differences

Ankylosing spondylitis (AS) is a chronic form of arthritis which limits the motion of the spine and muscles. A study was carried out at the Royal National Hospital for Rheumatic Diseases in Bath to see if daily stretching of tissues around the hip joints would help patients with AS to get more movement in their hip joints.

Thirty-nine consecutive admitted patients with 'typical' AS were allocated randomly to a control group receiving the standard treatment or to the treatment group receiving additional stretching exercises, in such a way that patients were twice as likely to be allocated to the 'stretched' group. The patients were assessed on admission and then three weeks later. For each patient several measurements were made on each hip, such as the extent of flexion, extension, abduction and rotation. This study is concerned just with flexion and lateral rotation, where all measurements are in degrees and an increase represents an improvement. The data are presented in Table B.13 and can be obtained from the author via computer e-mail as an ASCII file. No more details of the data-collection method will be given here. The statistician would normally analyse the data in collaboration with a physiotherapist, but here you are expected to use your common sense.

The question as posed by the hospital researchers was: 'Has the stretched group improved significantly more than the control group?' Your report should attempt to answer this question as well as to describe any other analysis which you think is appropriate and to discuss any queries which come to mind.

This is a fairly substantial data set which will take some time to analyse. As a rough guide, I suggest:

- thinking and looking time: 1 hour;
- preliminary examination of the data: 1–2 hours;
- further analyses as appropriate: 2 hours;
- writing up the report: 3 hours.

Table B.13 Measurements of flexion and rotation in degrees before and after treatment for 39 patients (or 78 hips). An odd-numbered row shows observations for the right hip of a particular patient and the next even-numbered row shows observations for the patient's corresponding left hip

Hip No.	Flexion		Rotation	
	Before	After	Before	After
(a) *Control group*				
1	100	100	23	17
2	105	103	18	12
3	114	115	21	24
4	115	116	28	27
5	123	126	25	29
6	126	121	26	27
7	105	110	35	33
8	105	102	33	24
9	120	123	25	30
10	123	118	22	27
11	95	96	20	36
12	112	120	26	30
13	108	113	27	30
14	111	109	15	14
15	108	111	26	25
16	81	111	14	13
17	114	121	22	24
18	112	120	26	26
19	103	110	36	41
20	105	111	33	36
21	113	118	32	35
22	112	115	27	31
23	116	120	36	30
24	113	121	4	2
(b) *Treatment group*				
1	125	126	25	36
2	120	127	35	37
3	135	135	28	40
4	135	135	24	34
5	100	113	26	30
6	110	115	24	26
7	122	123	22	42
8	122	125	24	37
9	124	126	29	29
10	124	135	28	31
11	113	120	22	38
12	122	120	12	34
13	130	138	30	35
14	105	130	30	27
15	123	127	33	42
16	125	129	34	40

Table B.13 continued

| Hip | Flexion | | Rotation | |
No.	Before	After	Before	After
17	123	128	18	27
18	126	128	26	32
19	115	120	20	40
20	125	120	22	40
21	120	135	7	20
22	105	127	10	28
23	120	130	25	28
24	110	120	25	26
25	127	135	35	39
26	125	130	25	29
27	128	138	30	43
28	124	136	33	45
29	115	124	26	34
30	116	124	40	42
31	106	110	20	21
32	111	116	16	17
33	113	114	10	12
34	105	115	11	14
35	100	105	14	23
36	99	88	2	10
37	113	119	48	50
38	126	126	35	35
39	102	110	22	30
40	94	115	13	24
41	116	111	22	25
42	112	114	20	25
43	122	128	30	35
44	122	128	33	35
45	118	127	22	30
46	115	124	25	27
47	115	125	32	39
48	118	127	12	25
49	78	121	35	34
50	77	126	30	32
51	129	129	36	44
52	127	132	33	25
53	127	127	10	15
54	132	139	39	36

Exercise B.9 Effect of anaesthetics – one way ANOVA

A study was carried out at a major London hospital to compare the effects of four different types of anaesthetic as used in major operations. Eighty patients undergoing a variety of operations were randomly assigned to one of the four anaesthetics and a variety of observations were taken on each patient both before and after the operation (see the Prelude). This exercise concentrates on just one of the response variables, namely the time, in minutes, from the reversal of the anaesthetic until the patient opened his/her eyes. The data are shown in Table B.15. Is there any evidence of differences between the effects of the four anaesthetics?

Table B.15 Time, in minutes, from reversal of anaesthetic till the eyes open for each of 20 patients treated by one of four anaesthetics (A–D)

A	B	C	D
3	6	3	4
2	4	5	8
1	1	2	2
4	1	4	3
3	6	2	2
2	2	1	3
10	1	6	6
12	10	13	2
12	1	1	3
3	1	1	4
19	1	1	8
1	2	4	5
4	10	1	10
5	2	1	2
1	2	1	0
1	2	8	10
7	2	1	2
5	1	2	3
1	3	4	9
12	7	0	1

Notes on Exercise B.1

This problem would have been more typical of both real life and statistics textbook questions if the reader were simply asked to carry out a test of significance. This would make the problem much easier in one way (the mechanics of a two-sample test are straightforward) but much harder in other ways.

Let us see what happens if we dive straight in to carry out the 'easy' significance test. There are two groups of observations to compare, but no natural pairing and so a two-sample t-test would appear to be appropriate. Let μ_C, μ_R denote the population mean concentrations for cut shoots and for rooted plants, and let $\bar{x}_C$, $\bar{x}_R$ denote the corresponding sample means. Then to test

$$H_0 : \mu_C = \mu_R$$

against

$$H_1 : \mu_C \neq \mu_R$$

we calculate

$$t_{obs} = (\bar{x}_C - \bar{x}_R)/s \bigg/ \sqrt{\left(\frac{1}{10} + \frac{1}{10}\right)}$$

which is distributed as t_{18} if H_0 is true, where s^2 denotes the combined estimate of within-group variance which is assumed to be the same in both groups. We find $\bar{x}_C = 46.7$, $\bar{x}_R = 36.5$, $s = 9.1$ and $t_{obs} = 2.51$ which is not quite significant at the 1% level for a two-tailed test. Thus we fail to reject H_0.

The above analysis treats the data mechanically. Despite its unsatisfactory nature, it is unfortunately what many people have been taught to do. Let us now return to the question and start, as suggested, by summarizing the data. It is helpful to order the two groups by size and calculate the means (or medians?) and standard deviations (or ranges?) of each group. Even the most cursory examination of the data reveals an obvious outlier in the first group, namely 18. This can be highlighted by drawing a pair of boxplots or simply tabulating the group frequencies as follows:

	Concentration								
	15–19	20–24	25–29	30–34	35–39	40–44	45–49	50–54	55–59
Cut shoots	1					1	3	3	2
Rooted plants			2	2	2	3	1		

The entire analysis depends crucially on what, if anything, we decide to do about the outlier. If possible, we should check back to see if it is an error. If the outlier is ignored, then a visual inspection suggests there is a significant difference, and this is confirmed by a revised t-value of 5.1 on 17 degrees of freedom. This is significant at the 1% level, giving strong evidence to reject H_0. Alternatively, the outlier may indicate a different-shaped distribution for cut shoots, in which case a t-test is not strictly appropriate anyway. I leave it to the reader to judge the best way to analyse and present the results. My own view is that there is evidence of a difference between means with or without the outlier.

Do any other questions occur to you? I suggest there are fundamental queries about the whole problem which are potentially more important than the analysis questions considered above. First, no objective is stated in the problem. Presumably we want to compare the chemical concentration in cut shoots with that in rooted plants. Is there any other background information? Have similar tests been carried out before? Why should the population means be exactly equal? Is it more important to estimate the difference in group means? Are the samples random, and, if so, from what populations? These questions should really be answered before we do anything! (So this exercise is a little unfair!)

Moral

Get background information before starting a statistical analysis. Note that even a single outlier can have a crucial effect on the results of the analysis.

Notes on Exercise B.2

There are five groups of observations to compare and so the 'standard' method of analysis is a one-way ANOVA. Any computer package should give an F-ratio of 15.3 on 4 and 40 degrees of freedom. This is significant at the 1% level, giving strong evidence that real differences exist between the group means. This could be followed by a least-significant-difference analysis or by a more sophisticated multiple comparisons procedure to show that group C achieves the best results and group E the worst. Note that these comparison procedures use the estimate of residual variance provided by the ANOVA, so that an ANOVA is not just used for testing.

However, you will understand the data better if the ANOVA is preceded by an IDA. The latter is essential anyway to check the assumptions on which the ANOVA is based. First calculate summary statistics for each group. With small equal-sized groups, the range may be used to measure spread.

Group	Mean	Range
A	19.7	10
B	18.3	10
C	27.4	7
D	23.4	9
E	16.1	11

The roughly constant within-group variation supports the homogeneous variance assumption of the ANOVA. For small samples, the standard error of

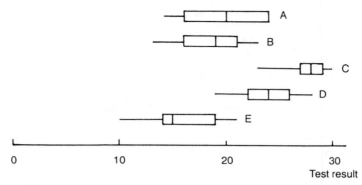

Figure B.1 Boxplots of test results for five groups of students.

a sample mean, namely $s/\sqrt{n}$, is approximately equal to range/n, which is about one in this case. This makes the differences in group means look relatively high. The differences become clearer in the set of boxplots in Fig. B.1. There is no overlap between groups C and E, while B and C only just 'touch'. It is arguable that no formal analysis is required to demonstrate that there really are differences between teaching methods.

If you still wish to calculate a P-value to confirm your subjective judgement, then there is nothing technically wrong with that here (although in general unnecessary statistical analyses should be avoided), but, whatever analysis is used, it is probably unwise to try and draw any general conclusions from such small artificial samples. Rather than concentrate on 'significance', it is more important to estimate the differences between group means (together with standard errors). If the differences are thought to be important from an educational point of view, then more data should be collected to ensure that the results generalize to other situations.

Moral

Summarize the data before performing significance tests. Checking whether results generalize is more important than significance testing anyway.

Solution to Exercise B.3

You should not rush into carrying out an ANOVA. As usual, your analysis should start with an IDA. Calculate the group means and plot them, as for example in Fig. B.2 where the lines connect group means and the 'whiskers' show the range of each sample of size four. Is your graph as clear? Did you remember to label the scales? Have you shown the within-group variation as well as the group means?

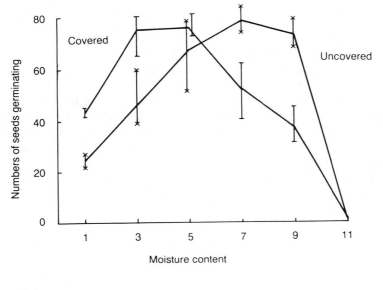

Figure B.2 Numbers of seeds germinating.

Looking at Fig. B.2, we see that numbers germinating increase to a maximum around level 7 for uncovered boxes, while for covered boxes the maximum is clearly at a lower level between 3 and 5. Judged against the relatively small within-group scatter, the differences between group means look 'large'. It is clear not only that water does have an effect, but also that covering a box has an effect. Only at level 5 do the results from covered and uncovered boxes overlap. The results at level 11 are different in that the seeds are swamped and the values are all zero in both groups.

Is there any point in confirming this assessment with formal tests of significance via an ANOVA? In this case I suggest the answer is no, even though that is what the biologist wants. This emphasizes that a statistician should not always answer the question which is posed but rather tackle the problem in the way thought to be most appropriate.

There are several reasons why hypothesis testing is inappropriate here. First, there is much prior information that water does affect plant growth, and it would be silly to ignore this and set up a null hypothesis that water has no effect. Second, the results of the significance tests are obvious beforehand from Fig. B.2, although the ANOVA will not tell us how the hypotheses are rejected. The analyst still has to draw something like Fig. B.2 to see where the maxima occur. Third, an ANOVA will be difficult to carry out anyway. Any reader who ploughed straight into an ANOVA will have spent unnecessary energy worrying about the missing observation even though it obviously makes no qualitative difference to the conclusions. In addition, Fig. B.2 is needed to see what secondary assumptions are reasonable. A standard ANOVA based on normal errors is clearly inappropriate since an assumption of constant variance cannot be made (compare the residual variation at levels 7 and 11). To a first approximation, the data may be regarded as binomial counts and it is possible to carry out an analysis of proportions, though it will not be fruitful. If the biologist insists on an ANOVA, perhaps because the results are to be published, then it may be argued that there is nothing 'wrong' in calculating P-values to confirm one's subjective judgement. However, this does not alter the fact that the null hypotheses would be silly and obviously rejected.

Some people may think it useful to fit a model, such as a pair of regression curves. However, in my view the presentation of Fig. B.2 is both necessary and sufficient and thus obviates the need for a more elaborate analysis.

Moral

Do not always do what you are asked to do!

Comments on Exercise B.4

The data in both tables consist of observed frequencies as opposed to measured variables. Such data are often called **count** data or **categorical** data, and the two-way tables of frequencies are often called **contingency tables**.

You should begin by looking at the tables. Various rules for improving the presentation of a table are given in section 6.5.2 of Part One. The only modification which needs to be made here is that row and column totals should be added to Table B.3 (as they already have in Table B.4). This gives Table B.5.

Table B.5 Observed frequencies of people with a particular combination of hair and eye colour

Eye colour	Hair colour				
	Black	Brunette	Red	Blond	Total
Brown	68	119	26	7	220
Blue	20	84	17	94	215
Hazel	15	54	14	10	93
Green	5	29	14	16	64
Total	108	286	71	127	592

Looking at the columns of Table B.5, we see that brown eyes are in the majority, except for the blond hair column where there are far more people

Table B.6 Average number of children per family in different income groups (counting '≥ 4' as '4')

Yearly income (1000 kr)	0–1	1–2	2–3	3+
Average number of children	0.89	0.94	0.76	0.60

with blue eyes. Indeed, the pattern in the blond column looks quite different to the rest. Alternatively, looking at the rows, we see that brown and blue eyes have roughly equal row totals, but give substantially different values in the black and blond columns. Clearly there is some association between hair and eye colour.

Applying a similar approach to Table B.4, we look at the columns and see that the highest frequency is one child for lower-income families but zero children for higher-income families. Perhaps poorer families tend to be larger. A crucial difference between Tables B.3 and B.4 is that whereas both variables in the former are nominal (since colours have no particular order), both variables in the latter are ordinal. This means that there may be alternative ways of summarizing the data. For example, the mean of the frequency distribution in each column of Table B.4 may be calculated as in Table B.6 by counting '≥ 4' as '4'. We can now see a little more clearly how family size relates to income, but the crude form of categorization used for income is less than ideal.

Are any inferential methods appropriate? Most students will have learnt to analyse contingency tables by means of a χ^2 goodness-of-fit test. This tests the null hypothesis that 'rows and columns are independent', or, to be more precise, that the probability of an observation falling in any particular column does not depend on which row that observation is in (and vice versa). Expected frequencies under this hypothesis may be calculated using the formula (row total) × (column total)/grand total. Then the χ^2 test statistic is given by

$$\chi^2 = \sum[(\text{observed} - \text{expected})^2/\text{expected}]$$

summed over all cells. The corresponding degree of freedom (DF) is (number of rows − 1) × (number of columns − 1). For Table B.3, we find $\chi^2 = 138.3$ on 9 DF, while for Table B.4 we find $\chi^2 = 568.6$ on 12 DF. Both these values are highly significant, leading to rejection of the independence hypothesis. But we really knew this already. With such large sample sizes, the χ^2-test is nearly always significant anyway, and it is more important to ask *how* the independence hypothesis is rejected and whether the deviations are of practical importance. The main benefit of the χ^2-test is to provide expected frequencies which can be compared by eye with the observed frequencies. Thus Table B.7 shows, for example, the excess of subjects with blue eyes and blond hair, and the shortfall of people with brown eyes and blond hair. A similar table derived from Table B.4 would not be particularly fruitful, and Table B.6 is more useful with these ordinal variables.

Table B.7 Observed and expected frequencies of subjects with particular hair and eye colours

Eye colour	Black Obs.	Black Exp.	Brunette Obs.	Brunette Exp.	Red Obs.	Red Exp.	Blond Obs.	Blond Exp.	Total
Brown	68	40	119	106	26	26	7	47	220
Blue	20	39	84	104	17	26	94	46	215
Hazel	15	17	54	45	14	11	10	20	93
Green	5	12	29	31	14	8	16	14	64
Total	108		286		71		127		592

A simple analysis along the above lines gives a good idea of the data using only a pocket calculator. More complicated analyses, such as those using log-linear modelling and correspondence analysis (e.g. Diaconis and Efron, 1985; Snee, 1974) may occasionally prove fruitful for the expert but are hardly necessary in the vast majority of practical cases. It may be more important to consult an appropriate subject specialist on any speculation raised by this analysis to see if the findings are confirmed by other studies or by alternative theory. A biologist may advise on hair/eye colour, while a demographer may be able to shed light on Swedish families in the 1930s.

Another important question is whether the given data are likely to be representative of some wider population, so that genuine inference can be made. The sample in Table B.3 was collected as part of a class project by a group of university students and should not perhaps be taken too seriously. However, the much larger sample in Table B.4 was taken from the 1936 Swedish census and is likely to be more representative.

Moral

In a two-way table of counts, it is often helpful to compare the observed frequencies by eye with the expected frequencies calculated assuming independence between rows and columns.

Solution to Exercise B.5

These data are unusual in two respects. Firstly, the study was stopped at 192 days, and so any observation given as 192 could in fact be any number greater than 192. Data like these are called **censored** data, and arise widely in reliability studies and in clinical trials. Secondly, some rats have died from causes other than cancer and this makes comparisons between the groups more difficult.

Table B.9 Numbers of rats developing cancer, surviving and dying prematurely of other causes

	Group II DX	Group IV DXP	Group I D	Group III DP
No. dying of cancer	4	11	13	19
No. surviving with cancer	3	0	3	0
No. surviving without cancer	7	6	0	0
No. of 'other' deaths	6	3	4	1
No. of rats in group	20	20	20	20

You may not have seen data like these before. But do not despair! Let us see how far we can get using common sense. With censored data, it is inappropriate to calculate some types of summary statistic, such as the mean and standard deviation, because the value 192 days (still alive) is quite different to 191 days (dead). If more than half the sample have died, it is possible to calculate median lifetime, but an alternative way to start looking at the data is to treat them as binary (e.g. cancer/no cancer) and find the numbers of rats in each group which (a) develop cancer, (b) survive until 192 days, and (c) die prematurely from causes other than cancer. These are given in Table B.9 with the columns reordered to demonstrate the effects more clearly.

First we note the high number of cancers. Clearly drug D does promote cancer although a control group, receiving no drugs, might have been desirable to confirm this (the extent of prior knowledge is unclear). Comparing groups I and III with II and IV, we see clear evidence that X does inhibit cancer in that the proportion of rats developing cancer goes down substantially (from 87% to 45%) while the proportion surviving until 192 days goes up substantially (from 7% to 40%). However, we note the worrying fact that five rats in group II died of other causes within the first five days of the trial. Could X have lethal side-effects? See also the three early non-cancerous deaths in group IV.

Comparing group II (DX) with group IV (DXP) and group I (D) with group III (DP), we see that drug P does seem to have an accelerating effect, while comparing group IV with group I suggests that X has a 'stronger effect than P'. We now have a 'feel' for the data.

It is also worth noting that the data look somewhat suspect in that too many survival times are repeated. For example, there are four 2s in group II, five 43s in group III, and three 106s in group IV. If this effect cannot be explained by chance, then some external effects must be playing a role or observations may not be taken every day (ask the experimenter!).

Is any further analysis indicated? You may want to know if the results are significant and this is not obvious just by looking at the data. It may also be helpful to fit a model to estimate the main effects and interaction of drugs X and P. Snell (1987, p. 153) shows how to fit a proportional hazards model

(Appendix A.15 and the last paragraph of Example E.4) which shows that the main effects are significant and allows survival functions to be estimated. With appropriate expertise available, such an analysis can be recommended. However, a descriptive analysis, as given above, may be adequate, and perhaps even superior for some purposes, when there are doubts about the data or when the client has limited statistical expertise (although the statistician may want to do a 'proper' analysis for himself). The descriptive analysis concentrates on understanding the data. 'Significance' is mainly important if little background knowledge is available and it is expensive to replicate the experiment, but for the given data it seems silly to set up unrealistic null hypotheses which ignore the prior information about the effects of the drugs.

There is further difficulty in analysing these data in that there are competing risks of death from cancer and from other causes. It is tempting, for example, to compare the proportion developing cancer in groups I and II, namely 16/20 and 7/20, by means of a two-sample test of proportions, which gives a significant result. However, there were several early deaths in group II which did not give time for cancer to develop. So should the proportions be 7/15 rather than 7/20 or what? Clearly there are traps for the unwary! Snell (1987) treats deaths from other causes as giving a censored value for which the post-mortem classification is ignored.

We sum up by saying that the three drugs do have the effects as suggested, that X inhibits cancer more than P accelerates it, but that X may have nasty side-effects.

Moral

When faced with an unfamiliar data set, use your statistical judgement to decide which summary statistics to calculate, but be prepared to consult an 'expert' in the particular area of statistics and/or area of application.

Notes on Exercise B.6

Some obvious queries which come to mind concerning the data are: (a) How was the sample selected, and do the data provide a fair representation of Hiroshima survivors? (b) How was the radiation dose assessed? Was it based solely on the person's position at the time of the explosion? (c) Is diagnosis of leukaemia and of other cancers perfect?

Taking the data at face value, we have a two-way table in which each 'cell' contains three integer counts. You will probably not have seen data exactly like these before; however, you should not give up but rather 'look' at the table using your common sense. We are interested in seeing if deaths are related to age and/or radiation dose. In other words, we are interested in the (marginal) effects of age and of radiation dose as well as in possible interactions between the two.

Table B.11 Number of deaths from leukaemia per 1000 survivors tabulated against age and radiation dose

Age in 1950	Radiation dose in rads						Overall
	0	1–9	10–49	50–99	100–199	200+	
5–14	0.45	0.2	—	1.4	7.1	15	0.92
15–24	—	0.2	1.1	1.2	3.3	7	0.88
25–34	0.45	0.8	—	—	2.2	10	0.96
35–44	—	—	0.4	1.7	2.3	12	0.69
45–54	1.64	1.0	0.8	—	2.0	12	1.60
55–64	0.56	—	1.3	4.6	3.5	14	1.25
65+	0.45	0.8	—	—	—	11	0.62
Overall	0.50	0.44	0.55	1.24	3.20	10.6	

Even a cursory glance at Table B.10 suggests that while deaths from all other cancers increase with age, deaths from leukaemia do not. Thus we need to treat leukaemia and 'all other cancers' separately. We also need to look at proportions rather than the number of deaths in order to get fair comparisons. When proportions are small (as here), it is better to compute rates of death rather than proportions to avoid lots of decimal places. The number of deaths per 1000 survivors is a suitable rate which is $10^3 \times$ corresponding proportion. The resulting two-way table of rates for leukaemia is shown in Table B.11. A similar type of table may be obtained for 'other cancers'.

Constructing a nice clear two-way table is not trivial (see section 6.5.2 of Part One). Were your tables as clear? There are several points to note. First, the overall death rates are obtained as a ratio of the total frequencies, and not as the average of the rates in the table. Second, the overall row rates are given on the right of the table as is more usual. Third, overall column rates should also be given. Fourth, zero rates are indicated by a dash. Fifth, different accuracy is appropriate in different parts of a table as the sample sizes vary.

It is now clear that leukaemia death rates are strongly affected by radiation dose but little affected by age. In contrast, 'other cancers' are affected by age but not by radiation dose. There is little sign of interaction between age and radiation dose or of any outliers when one bears in mind that the number of deaths for a particular age/dose combination is likely to be a Poisson variable and that one death may produce a relatively large change in death rate. For example, the apparently large leukaemia death rate for age 55–64 and dose 50–99 rads is based on only two deaths.

The above findings are so clear that there seems little point in carrying out any significance tests. What could be useful is to fit some sort of curve to describe the relationship between, say, leukaemia death rate and radiation dose. Alternatively, we could carry out a more formal inferential analysis which could here involve fitting some sort of log-linear or logistic response model (see

Exercise G.5 for an example of a logistic model). Such a model might provide a proper probabilistic foundation for the analysis and be well worthwhile for the expert analyst. However, the analysis presented here is much simpler, will be understood by most analysts and clients, and may well be judged adequate for many purposes, particularly if there are doubts about the reliability of the data.

Moral

Constructing a clear two-way table is not trivial, particularly when the 'cell' values are ratios of two frequencies.

Notes on Exercise B.7

The statistician has to cope with samples ranging from thousands to single figures. Here the sample sizes are very small and the power of any test will be low. Thus we cannot expect the treatment effect to be significant unless the effect is very large. Rather we can treat the data as a pilot study to see if the vaccine is worth testing on a larger sample in a variety of conditions. Thus, rather than ask 'Is the treatment effect significant?' – though it will be a bonus if it is – it is more sensible to ask 'Is there something interesting here?'.

If we were to perform a two-sample test on the two group means, we would find that the result is nowhere near significant even though the two sample means, 24.1 and 14.8, differ substantially. Yet a glance at Table B.12 suggests that some observations in the treatment group are much lower than those in the control group and that there is an effect of potential practical importance here. Let us look at the data in more detail using an IDA. There are too few observations to plot histograms, and so we plot the individual values as a pair of dotplots in Fig. B.3. In order to distinguish between the two samples, one dot plot is plotted using crosses. It is now clear that the shape of the distribution of worm counts for the treatment group is quite different to that of the control group. Four lambs appear unaffected by the vaccine, while four others have given large reductions. The group variances are apparently unequal and so one assumption for the usual t-test is invalid. If we carry out a two-sample Welch test, which allows for unequal variances, the result is much closer to being

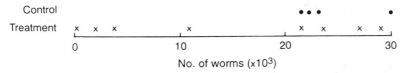

Figure B.3 Worm counts for four control lambs and eight vaccinated lambs.

significant, but this or any other parametric approach is heavily dependent on the assumptions made. A non-parametric approach could be affected by the fact that there may be a change in shape as well as in the mean. Yet the IDA, in the form of Fig. B.3, suggests that there is something interesting here, namely that the vaccine works on some lambs but not on others. This possibility is supported by knowledge about the behaviour of other vaccines which can have different effects on different people. Once again it is vital to use background information. Unless other experimental vaccines do better in these preliminary trials, it appears worth taking larger samples for this treatment. There seems little point in any further analysis of this small data set.

Moral

When samples are very small, look at the data to see if the systematic effects are of potential practical importance, rather than rushing into using a technique which depends on dubious assumptions.

Discussion of Exercise B.8

I have no definitive analysis for these data, but various sensible suggestions can be made.

The first thing many students will try is to compute the change for each hip and then carry out a two-sample t-test to compare the changes in the treatment group with the changes in the control. When this is done, the rotation in the treatment group is found to improve significantly more than in the control group ($t = 3.8$ is significant at the 1% level) but that the results for flexion ($t = 1.7$) are only just significant at the 5% level if a one-tailed test is thought appropriate. But is this analysis appropriate? This book should have taught you not to rush into significance testing.

The first problem is choosing a sensible response variable. Should (absolute) changes be analysed or perhaps percentage change? It could be argued that a given absolute change in score is more beneficial for a severely handicapped patient than for a more mobile patient. However, there is an upper bound to the scores (180°?) which will affect percentage change more than absolute change. In any case, it is possible for changes to go negative and one hip decreases from 4° to 2° giving a misleading reduction of 50%. So perhaps we have to stick to absolute changes even though they are not ideal.

Before doing any analysis, we should first look at the data by means of an IDA. Do the data look reliable? Are there any outliers? Are the treatment and control groups comparable? If we look at the distribution of the final recorded digits we see that there are 'too many' zeros and fives. Measuring how much a hip will bend is clearly difficult to perform accurately and some rounding is perhaps inevitable; but this should not affect the conclusions too much. If

Table B.14 Mean values for the hip data

	Before	After	Change
(a) *Flexion*			
Control	110.0	113.7	3.7
Treatment	116.5	124.0	7.5
(b) *Rotation*			
Control	25.0	26.0	1.0
Treatment	24.8	31.4	6.6

stem-and-leaf plots of starting values and of changes are formed, a number of potential outliers can be seen (e.g. the improvement in flexion for treatment hips 49 and 50). There is no obvious reason to exclude any observation but these outliers do negate a normality assumption. Summary statistics should be calculated and the mean values are given in Table B.14. (Query: should medians be preferred here?) Note that the treatment group produces higher changes, particularly for rotation. Boxplots of improvements for each of the different groups are also helpful in displaying the apparent benefit of the treatment.

There is some evidence that the control group has a lower initial mean for flexion, which raises the question as to whether a random allocation procedure really was used. Then a relevant question is whether change is related to the corresponding initial score. In fact these two variables are negatively correlated (down to -0.6 for the control flexion group), indicating that severely handicapped patients tend to improve more than others, although these correlations are inflated by outliers. The correlation suggests that an analysis of covariance may be appropriate. If we ignore the initial information, then we are being 'kind' to the control group and this may help to explain why the flexion results are not significant in the above-mentioned t-test.

Another problem with these data is that there are measurements on two hips for each patient. Analysis shows that observations on the two hips of a patient are positively correlated (up to over 0.7 in one group). Thus observations are not independent, and our effective total sample size is between 39 and 78, rather than 78 as assumed in the earlier t-test. There is also a small correlation between improvements in flexion and rotation for the two hips of a particular patient. We could average each pair of observations, giving 39 independent observations, but this loses information about between-hip variation. An alternative is to analyse right and left hips separately. The effect of ignoring the correlation in an overall t-test of all individual hips is to overestimate the extent of significance. It may be possible to allow for 'patient' effects in a more complicated analysis, but it is not clear that this is worth the effort in an exploratory study like this.

A completely different approach is to use non-parametric methods. The simplest approach is to see what proportion of hips improved in the control and in the treatment groups. The results are shown below. Note that the treatment group gives better results, particularly for rotation.

	Flexion	Rotation
Control	0.75	0.54
Treatment	0.83	0.89

The difference in proportions is significant for rotation but not for flexion. Alternatively, we can carry out a more powerful non-parametric test on the **ranks**, called a two-sample Mann–Whitney test. This suggests that the rotation results are significant at the 1% level (as for the t-test) and that the flexion results are significant at the 5% level. This test still ignores the problems due to correlated pairs and to unequal initial flexion scores. However, these two effects will tend to cancel each other out. The test does not rely on a normality assumption and, in view of the outliers, seems a better bet than the t-test.

Hitherto we have considered flexion and rotation separately. A more thorough and sophisticated analysis would consider them together. For example, does a patient whose rotation responds well to treatment also have a good flexion response? Is it possible to form a single variable from the measured responses which provides an overall measure of patient response? We will not pursue this here.

The overall conclusion is that the new treatment does seem to work, especially for improving rotation. However, we should also note a number of queries such as: (a) Were patients really allocated randomly? (b) What is a 'typical' AS patient? (c) How expensive is the treatment and will its effect last?

Moral

The problems encountered here, such as non-normality and correlations between observations are 'typical' of real data, and so the analyst must always be on the lookout for departures from 'standard' assumptions.

Notes on Exercise B.9

By concentrating on one variable, and ignoring other information, the following analysis should be regarded as exploratory. It will at least give some idea as to whether there are 'large' differences between the effects of the anaesthetics. It is usually a good idea to start by looking at a large set of variables one or two at a time, before moving on to look at the multivariate picture.

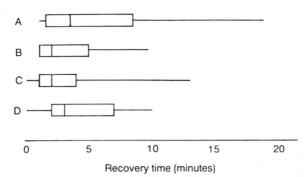

Recovery time (minutes)

Figure B.4 Boxplots of recovery times for groups of 20 patients treated by one of four anaesthetics (A, B, C or D).

It is easy to reorder the four groups of observations by size and then construct boxplots, as in Fig. B.4. In Exercise B.2 we were also interested in comparing observations in several groups and there a set of boxplots revealed obvious differences between groups. Here the set of boxplots shows substantial overlap between groups and suggests that there are not significant differences between groups. The plots also reveal that the distributions are severely skewed, which suggests that medians are better measures of location than means. Some descriptive statistics are shown in Table B.16.

The differences between group means (or medians) are small compared with the variability within groups. The obvious inferential technique for comparing several groups is the one-way ANOVA. However, this technique assumes that observations are normally distributed. This is not true here. We could ignore the problem and rely on the robustness of ANOVA, or use a non-parametric approach, or transform the data to normality. We try the latter. If a Box–Cox transformation is applied (see section 6.8), the power parameter λ could be estimated by maximum likelihood, but it is easier in practice to use trial and error as only a few special transformations make sense. Trying square roots, cube roots and logarithms, it is found that logarithms give a roughly symmetric distribution in each group. As there are some zero survival times, and log 0 is infinite, the analyst needs to use $\log(x + k)$ rather than log x. By trial and error, $k = 1$ is found to be a reasonable value. It is this sort of 'trick' which is so necessary in practice but which is often not covered in textbooks.

Table B.16 Some summary statistics on groups A–D of Fig. B.4

	A	B	C	D
Mean	5.4	3.2	3.0	4.3
Median	3.5	2.0	2.0	3.0
Range	1–19	1–10	0–13	0–10

Table B.17 One-way ANOVA

Source	SS	DF	MS	F
Anaesthetics	2.28	3	0.76	1.8
Residual	32.50	76	0.43	
Total	34.78	79		

A one-way ANOVA of the transformed data produces results shown in Table B.17. Note that the F-value only needs to be given to one decimal place.

The F-ratio is nowhere near significant at the 5% level and so we can accept the null hypothesis that there is no difference between the four anaesthetics in regard to recovery time. It is interesting to note that if an ANOVA is carried out on the raw data, then, by a fluke, the F-ratio happens to take exactly the same value. This is an indication of the robustness of ANOVA.

Is any follow-up analysis indicated? The estimated residual mean square, namely 0.43, could be used to calculate confidence intervals for the means of the transformed variables, and it is worth noting that recovery times for anaesthetic A do seem to be a little longer. It would be interesting to see if this effect recurs in further cases.

Moral

Be prepared to transform or otherwise modify your data before carrying out a formal analysis, but note that ANOVA is robust to moderate departures from normality.

C
Correlation and regression

When observations are taken simultaneously on two or more variables, there are several ways of examining the relationship, if any, between the variables. For example, principal component analysis (or even factor analysis?) may be appropriate if there are several variables which arise 'on an equal footing'. In this section we consider two rather simpler approaches. A **regression** relationship may be appropriate when there is a response variable and one or more explanatory variables. A **correlation** coefficient provides a measure of the linear association between two variables. These techniques are straightforward in principle. However, the examples in this section demonstrate the difficulties which may arise in interpreting measures of correlation and in fitting regression relationships to non-experimental data.

Exercise C.1 Correlation and regression – I

This is a simple exercise, but with an important message. Figure C.1 shows some observations on two chemical variables arising in a chemical experiment. The vertical axis is a partition coefficient while the horizontal axis is the volume fraction of one of two chemicals in a mixed solvent, but the full chemical details are unnecessary to what follows. The task is simply to make an educated guess as to the size of the correlation coefficient without carrying out any calculations. Note that the straight line joining the 'first' and 'last' points is not the least-squares line but was inserted by the chemist who produced the graph.

Exercise C.2 Correlation and regression – II

This is a 'fun' tutorial exercise on artificial data which nevertheless makes some valuable points. Figure C.2 shows four sets of data. In each case, comment on the data, guesstimate the value of the correlation coefficient, and say whether you think the correlation is meaningful or helpful. Also comment on the possibility of fitting a regression relationship to predict y from x.

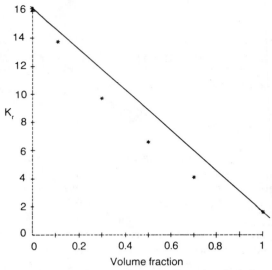

Figure C.1 Observations on two variables for the chemical hexane-benzonitrile-dimesulfoxide.

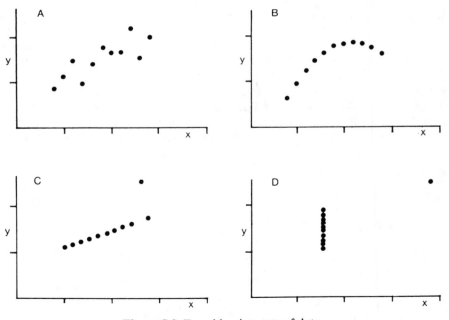

Figure C.2 Four bivariate sets of data.

Exercise C.3 Sales data – multiple regression?

Table C.1 shows the sales, average price per ton, and advertising support for a certain commodity over six years. Find the regression relationship between sales, price and advertising and comment on the resulting equation. Does the equation describe the effects of different price and advertising levels on sales? Can you think of an alternative way of summarizing the data?

Table C.1 Sales, price and advertising data

	1979	1980	1981	1982	1983	1984	Average
Sales (in £million), S	250	340	300	200	290	360	290
Price (in £), P	25	48	44	20	38	60	39
Advertising (in £'000), A	35	32	38	30	34	46	36

Exercise C.4 Petrol consumption – multiple regression

Table C.2 shows the petrol consumption, number of cylinders, horse power, weight and transmission type (automatic or manual) for 15 American car models made in 1974. This is part of a data set formed by Dr R. R. Hocking which was reanalysed by Henderson and Velleman (1981). The original data set

Table C.2 The petrol consumption (in miles per US gallon), number of cylinders, horse power, weight (in '000 lb) and transmission type for 15 1974-model American cars

Automobile	MPG	No. of cylinders	HP	Weight	Transmission
Mazda RX-4	21.0	6	110	2.62	M
Datsun 710	22.8	4	93	2.32	M
Hornet Sportabout	18.7	8	175	3.44	A
Valiant	18.1	6	105	3.46	A
Duster 360	14.3	8	245	3.57	A
Mercedes 240D	24.4	4	62	3.19	A
Mercedes 450SLC	15.2	8	180	3.78	A
Cadillac Fleetwood	10.4	8	205	5.25	A
Lincoln Continental	10.4	8	215	5.42	A
Fiat 128	32.4	4	66	2.20	M
Toyota Corolla	33.9	4	65	1.84	M
Pontiac Firebird	19.2	8	175	3.84	A
Porsche 914-2	26.0	4	91	2.14	M
Ferrari Dino 1973	19.7	6	175	2.77	M
Volvo 142E	21.4	4	109	2.78	M

contained more variables and a greater variety of car models, but this data set is quite large enough to get experience with. (Though normally the larger the sample the better!) The object of this exercise is stated as being to find a regression equation to predict petrol consumption (measured as miles per US gallon under specified conditions) in terms of the other given variables.

Solution to and discussion of Exercise C.1

If, like me, you guessed a value of about -0.8 to -0.9, then you are wide of the mark. The actual value is -0.985. Showing this graph to a large number of staff and students, the guesses ranged widely from -0.2 to -0.92 (excluding those who forgot the minus sign!), so that the range of guesses, though wide, did not include the true value.

Experienced statisticians know that correlation can be difficult to guesstimate (Cleveland, Diaconis and McGill, 1982). Our judgement can be affected by the choice of scales, the way the points are plotted and so on. One reason why everyone underestimated the true value in this case could be that the straight line exhibited in the graph is not a good fit. A second reason is that the eye can see that the departures from linearity are of a systematic rather than random nature. However, the main reason arises from the (unexpected?) nature of the relationship between correlation and the residual standard deviation, which is

$$s_{y|x}^2 = \text{residual variance of } y = s_y^2(1 - r^2)$$

where s_y is the (unconditional) standard deviation of y, and r is the correlation. When $r = -0.985$, we have $s_{y|x} = 0.17 \, s_y$, so that, despite the high correlation, the residual standard deviation for a linear model would still be 17% of the unconditional standard deviation. The measure $1 - s_{y|x}/s_y = 1 - \sqrt{(1 - r^2)}$ comes closer to most people's perception of correlation. You are invited to guess and then evaluate r when (a) $s_{y|x} = 0.5 s_y$ and (b) $s_{y|x} = 0.7 s_y$.

The data were given to me by a colleague in the Chemistry Department, where there is controversy regarding this type of relationship. One school of thought says: 'The very high correlation shows the relationship is linear'. A second school of thought says: 'Despite the very high correlation, the departures from linearity are systematic so that the relationship is not linear'. I hope the reader can see that the argument used by the first school of thought is false. I supported the alternative view having seen several similar non-linear graphs and having heard that the residual standard deviation is known to be much smaller than the residual standard deviation for a linear model. It is of course potentially misleading to calculate correlations for data exhibiting obvious non-linearity, and if you refused to do so in this case, then you are probably wise!

Indeed your first reaction to this exercise may (should?) have been to query whether it is sensible to try and guess a preliminary estimate for the correlation or even to be interested in the correlation at all. Nevertheless, the statistician has to be prepared to understand and interpret correlations calculated by other scientists.

Another striking example, demonstrating that an apparently high correlation (0.992!) can be misleading, is given by Green and Chatfield (1977, p. 205). In their example, the correlation is grossly inflated by an extreme outlier.

Moral

Correlations are difficult to assess and interpret.

Solution to Exercise C.2

These data were constructed by Anscombe (1973), so as to have an *identical* correlation coefficient, namely 0.82, even though the data sets are very different in character. (These data are also discussed by Weisberg, 1985, Example 5.1.)

Data set A looks roughly bivariate normal and the correlation is meaningful. Set B is curvilinear and, as correlation is a measure of linear association, the correlation is potentially misleading. If, say, a quadratic model is fitted, then the coefficient of determination is likely to be much higher than 0.82^2. Data set C lies almost on an exact straight line except for one observation which looks like an outlier. The correlation only tells part of the story. Set D looks very unusual. The x-values are all identical except for one. The latter is a very influential observation. With only two x-values represented, there is no way of knowing if the relationship is linear, non-linear, or what. No statistics calculated from these data will be reliable.

As to a possible regression relationship, the construction of the data is even more cunning in that corresponding first and second moments are all identical for each data set. Thus the fitted linear regressions will be identical, as well as the correlations. However, although the fitted lines are identical, an inspection of the residuals should make it apparent that the linear model for data set A is the only one that can be justified for the given data.

Moral

Look at the scatter diagram before calculating regression lines and correlations.

Discussion of Exercise C.3

The first part of the question is an 'easy' piece of technique. Using MINITAB or some similar package, a multiple regression equation can easily be fitted to the data giving

$$S = 4.26P - 1.48A + 176 + \text{error}$$

with a coefficient of determination given by $R^2 = 0.955$. Is this equation useful and does it describe the effects of different price and advertising levels on sales? The short answer is no.

First you should query whether it is wise to fit a three-parameter model to only six observations. In particular, the standard errors of the fitted coefficients are relatively large. Far more observations are desirable to get an empirically-based model which describes behaviour over a range of conditions. Indeed, it could be argued that this is too small a data set to be worth talking about, but I have included this exercise because, rightly or wrongly, people do try to fit equations to data sets like this. Thus the statistician needs to understand the severe limitations of the resulting model.

This data set is also ideal for demonstrating the problem caused by correlated explanatory variables. A second obvious query about the fitted model is that the coefficient of A is negative. However, if sales are regressed on advertising alone, then the coefficient of A does turn out to be positive as one would expect. Thus the introduction of the second variable, P, not only alters the coefficient of A but actually changes its sign. This emphasizes that the analyst should not try to interpret individual coefficients in a multiple regression equation except when the explanatory variables are orthogonal.

Rather than start with a technique (regression), it would be better as usual to start with an IDA and also ask questions to get background information. In particular, you should have plotted the data, not only S against P and against A, but also A against P. From the latter graph the positive correlation of the explanatory variables is evident. Advertising was kept relatively constant over the years 1979–83 and then increased sharply in 1984 when price also increased sharply. We cannot therefore expect to be able to separate satisfactorily the effect of advertising from that of price for the given set of data. The fitted equation is simply the best fit for the given set of data and does not describe the effects of different price and advertising levels on sales. In particular, if price were to be held constant and advertising increased, then we would expect sales to increase even though the fitted model 'predicts' a decrease. It is possible to find an alternative model with a positive coefficient for advertising for which the residual sum of squares is not much larger than that for the least-squares regression equation. This would be intuitively more acceptable but cannot be regarded as satisfactory until we know more about marketing policy.

An alternative possibility is to omit either price or advertising from the fitted model. A linear regression on price alone gives

$$S = 140 + 3.84P$$

with $R^2 = 0.946$, whereas the regression on advertising alone gives a much lower R^2, namely 0.437. Thus it seems preferable to 'drop' advertising rather than price to get a model with one explanatory variable.

The regression on price alone is one alternative way of summarizing the data. Is there another? The perceptive reader may suggest looking at the data in an entirely different way by considering a new response variable, namely volume = sales/price. The problem with the variable 'sales' when expressed in monetary terms is that it already effectively involves the explanatory variable price. The response variable 'volume' is arguably more meaningful than sales and leads to a locally linear relationship with price where volume decreases as price increases. This is to be expected intuitively. The linear regression equation for $V = S/P$ is given by

$$V = 12.2 - 0.11P$$

with $R^2 = 0.932$. Thus the 'fit' is not quite so good as for the regression of S on P, but does provide an alternative, potentially useful way of summarizing the data.

All the above regression equations give a 'good' or 'best' fit to the particular given set of data. However, as noted earlier, it is open to discussion whether it is wise or safe to derive a regression equation based only on six observations. There is no reason why the models should apply in the future to other sets of data, particularly if marketing policy is changed. In particular, a model which omits the effect of advertising may rightly be judged incomplete. The mechanical deletion of variables in multiple regression has many dangers. Here it seems desirable to get more and better data, to ask questions and exploit prior information to construct a model which does include other variables such as advertising. For example, advertising budgets are often determined as a fixed percentage of expected sales, but could alternatively be increased because of a feared drop in sales. Knowledge of marketing policy is much more important than 'getting a good fit'.

Moral

Multiple regression equations fitted to non-orthogonal, non-experimental data sets have severe limitations and are potentially misleading. Background knowledge should be incorporated.

Notes on Exercise C.4

Here miles per gallon is the response variable and the other four variables are predictor variables. This is a typical multiple regression problem in that the analyst soon finds that it is not obvious how to choose the form of the regression equation. Here the particular problems include correlations between the predictor variables and the fact that there is a mixture of discrete and continuous variables. This is the sort of data set where different statisticians may, quite reasonably, end up with different models. Although such models may appear dissimilar, they may well give a rather similar level of fit and forecasting ability. This data set has already given rise to different views in the literature and I shall not attempt a definitive solution.

Some general questions are as follows:

1. What is the correct form for a suitable model? Should non-linear or interaction terms be included, and/or should any of the variables be transformed either to achieve linearity or to reduce skewness? Which of the predictor variables can be omitted because they do not have a significant effect on the response variable? Is there background information on the form of the model – for example, that some predictor variables must be included?
2. What error assumptions are reasonable? What, for example, can be done about non-normality or non-constant variance if such effects occur?
3. Are there any errors or outliers which need to be isolated for further study or even removed from the data set? Which observations are influential?
4. What are the limitations of the fitted model? Under what conditions can it be used for prediction?

Cox and Snell (1981, Example G) discuss the problems in more detail and illustrate in particular the use of the logarithmic transformation and the fitting of interaction terms.

Returning to the given data set, start by plotting histograms (or stem-and-leaf plots) of the five (marginal) univariate distributions to assess shape and possible outliers. Then look at the scatter diagrams of the response variable with each of the four predictor variables to assess the form of the relationship (if any) and to see if a transformation is needed to achieve linearity. It is also a good idea to look at the scatter diagrams of each pair of predictor variables to see if they are themselves correlated. If some of the correlations are high then some of the predictor variables are probably redundant. This means looking at 10 ($= {}^5C_2$) graphs but is well worth the effort. Then a multiple regression model can be fitted to the data using an appropriate variable selection technique. You may want to try both forward and backward selection to see if you get the same subset of predictor variables. Having fitted a model, plot the residuals in whatever way seems appropriate to check the adequacy of the model, and modify it if necessary.

The interactive approach recommended here contrasts with the automatic approach adopted by many analysts when they simply entrust their data to a multiple regression program. This may or may not give sensible results. An automatic approach is sometimes regarded as being more 'objective', but bear in mind that the analyst has made the major subjective decision to abdicate all responsibility to a computer.

One regression program fitted an equation which included horse power and weight but excluded the number of cylinders and transmission type. As an example of an interesting residual plot, Fig. C.3 shows the resulting residuals plotted against one of the discarded predictor variables. There are two residuals

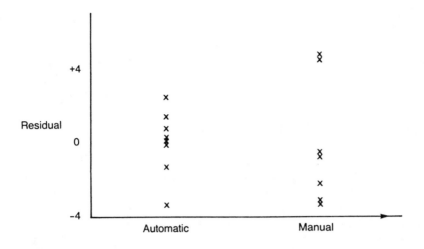

Figure C.3 An example of a residual plot.

which might be regarded as outlying which both give high values of miles per gallon for a manual transmission: the Fiat 128 and the Toyota Corolla. On looking back at Table C.2, we note that these two models have much higher values for miles per gallon than other models and the residual plot helps us to pick them out. There is no reason to think they are errors, but rather they are 'exceptional' data points which do not conform to the same pattern as the other cars.

It is mainly the responsibility of the automobile engineer to decide on the limitations of the fitted model and to assess whether or not it can be employed usefully to predict petrol consumption for a new model. It would be desirable to have a wider variety of cars to fit the model to, and in any case the deviations from the model are just as interesting as the predicted values.

Moral

Fitting a multiple regression model requires skilled interaction with the data and with an appropriate subject specialist.

D
Analysing complex large-scale data sets

The exercises in this chapter are concerned with more substantial multivariate data sets. They should give the reader valuable experience in handling a large-scale 'messy' set of data, and emphasize the point that a simple 'common-sense' approach to data analysis is often preferable to the use of more sophisticated statistical techniques.

With a large data set, it is easy to feel rather overwhelmed. Then it is tempting to feed the data into a computer, without checking it at all, and then apply some sophisticated analysis which may, or may not, be appropriate. The ostensible reason for this is that data scrutiny and IDA are too time-consuming for large data sets. In fact data scrutiny and data summarization are perhaps even more important for large data sets, where there are likely to be more errors in the data and more chance of going horribly wrong. The time involved will still be relatively small compared with the time spent collecting the data and getting it onto a computer.

Data scrutiny can be done partly 'by eye' and partly 'by computer'. As the size of the data set increases, it soon becomes impractical to look at all the tabulated raw data by eye, although I still recommend 'eyeballing' a small portion of the data. Nevertheless, the vast majority of the data scrutiny will be done with the aid of a computer, albeit backed up by an experienced eye. For example, the histogram of each variable can be quickly plotted by computer and then checked by eye for gross outliers.

There is increasing use of multivariate techniques such as principal component analysis and multi-dimensional scaling (section 6.6). Examples demonstrating the use of such techniques may be found in many books such as Gnanadesikan (1977), Chatfield and Collins (1980) and Everitt and Dunn (1991), and will not be duplicated here. These techniques can sometimes be very useful for reducing dimensionality and for giving two-dimensional plots. However, traditional descriptive statistics is still a valuable prerequisite, even when the data sets are larger than those considered here, and proves sufficient in the two examples considered here.

Table D.1 Code numbers, weights and concentrations of 14 chemicals for fodder rate at various stages of growth

Code No.	Weight (g)	Concentration (µg/g)														Stage of growth
00000100	0.0047	43.4	581.8	86.8	26.1	347.7	261.4	0.5	0.5	0.5	583.9	362.4	381.2	0.5	0.5	Stored seed
00000200	0.0038	131.6	526.3	26.3	67.4	811.1	315.7	0.5	0.5	0.5	281.2	227.4	628.9	0.5	0.5	
00000300	0.0035	137.6	1904.7	126.9	42.3	529.1	179.9	0.5	0.5	0.5	410.9	542.6	1000.3	0.5	0.5	
00000400	0.0040	346.2	650.2	52.1	72.9	610.2	412.7	0.5	0.5	0.5	500.1	195.2	286.1	0.5	0.5	
00000500	0.0027	30.2	695.4	105.8	15.1	451.4	752.6	0.5	0.5	0.5	110.9	241.7	471.9	0.5	0.5	
00000600	0.0019	22.4	111.9	370.3	179.2	246.4	425.5	0.5	0.5	0.5	171.4	342.1	311.6	0.5	0.5	
00000700	0.0028	441.2	142.0	168.1	63.0	681.7	479.6	0.5	0.5	0.5	353.6	141.9	871.2	0.5	0.5	
00000800	0.0038	92.9	1346.7	294.1	123.8	753.3	771.4	0.5	0.5	0.5	192.8	488.6	1126.2	0.5	0.5	
00000900	0.0045	288.9	2000.0	122.2	33.3	771.9	311.1	0.5	0.5	0.5	641.9	417.1	362.4	0.5	0.5	
00001000	0.0024	312.6	472.5	199.2	54.3	802.9	253.6	0.5	0.5	0.5	554.1	181.1	320.8	0.5	0.5	—
00001100	0.0078	91.6	109.9	16.8	0.5	695.0	531.1	0.5	0.5	0.5	622.7	494.5	527.6	0.5	0.5	1 day old plants
00001200	0.0057	150.3	62.6	15.8	0.5	676.7	451.1	0.5	0.5	0.5	576.3	526.3	499.8	0.5	0.5	
00001300	0.0103	375.4	540.8	38.8	0.5	841.4	673.1	0.5	0.5	0.5	504.8	427.2	431.6	0.5	0.5	
00001400	0.0047	171.2	60.8	13.2	0.5	152.0	243.2	0.5	0.5	0.5	152.0	121.6	133.4	0.5	0.5	
00001500	0.0079	153.2	306.5	46.6	0.5	666.2	253.2	0.5	0.5	0.5	273.1	259.8	262.4	0.5	0.5	
00001600	0.0055	77.9	597.4	28.9	0.5	1064.9	207.8	0.5	0.5	0.5	415.6	337.6	312.2	0.5	0.5	
00001700	0.0099	153.2	696.6	48.7	0.5	884.7	285.6	0.5	0.5	0.5	487.6	557.3	492.8	0.5	0.5	
00001800	0.0070	224.5	795.9	28.4	0.5	306.1	367.3	0.5	0.5	0.5	122.4	183.6	161.6	0.5	0.5	
00001900	0.0049	181.4	272.1	12.2	0.5	260.1	236.4	0.5	0.5	0.5	158.7	226.7	234.3	0.5	0.5	
00002000	0.0076	131.6	838.6	16.4	0.5	476.9	444.1	0.5	0.5	0.5	230.3	299.4	287.6	0.5	0.5	—
00002100	0.0113	99.4	81.5	0.5	0.5	541.4	262.3	0.5	0.5	0.5	241.2	184.9	87.4	0.5	0.5	2 days
00002200	0.0151	78.2	90.2	0.5	0.5	330.9	184.7	0.5	0.5	0.5	177.0	138.6	61.5	0.5	0.5	
00002300	0.0184	96.1	62.6	0.5	0.5	527.0	197.6	0.5	0.5	0.5	187.7	148.2	65.9	0.5	0.5	
00002400	0.0109	121.2	54.1	0.5	0.5	389.2	411.4	0.5	0.5	0.5	255.8	255.8	137.1	0.5	0.5	
00002500	0.0155	58.6	183.8	0.5	0.5	391.0	215.0	0.5	0.5	0.5	132.9	136.8	71.7	0.5	0.5	
00002600	0.0113	144.3	106.3	0.5	0.5	579.1	304.1	0.5	0.5	0.5	194.0	6.0	163.4	0.5	0.5	
00002700	0.0147	52.9	76.2	0.5	0.5	220.6	88.3	0.5	0.5	0.5	139.7	139.7	22.0	0.5	0.5	
00002800	0.0081	81.4	129.2	0.5	0.5	351.7	134.7	0.5	0.5	0.5	142.2	144.3	112.2	0.5	0.5	
00002900	0.0093	91.2	39.1	0.5	0.5	1303.3	221.6	0.5	0.5	0.5	280.2	267.2	73.9	0.5	0.5	
00003000	0.0107	199.0	161.7	0.5	0.5	971.9	934.6	0.5	0.5	0.5	336.4	186.9	29.9	0.5	0.5	
00003100	0.0298	64.0	10.7	0.5	0.5	152.5	79.3	0.5	0.5	0.5	68.6	56.4	71.2	0.5	0.5	
00003200	0.0270	76.4	8.4	0.5	0.5	213.5	96.7	0.5	0.5	0.5	270.3	270.3	270.3	0.5	0.5	
00003300	0.0236	−1.0	−1.0	−1.0	−1.0	−1.0	−1.0	−1.0	−1.0	−1.0	−1.0	−1.0	−1.0	−1.0	−1.0	—
	−1.0	−1.0	−1.0	−1.0	−1.0	−1.0	−1.0	−1.0	−1.0	−1.0	−1.0	−1.0	−1.0	−1.0	−1.0	

ID														
00003400	0.0280	95.9	8.2	0.5	0.5	214.3	95.9	0.5	0.5	65.3	55.1	64.3	0.5	0.5
00003500	0.0243	26.8	6.5	0.5	0.5	70.8	95.7	0.5	0.5	57.4	59.3	71.4	0.5	0.5
00003600	0.0176	27.8	13.9	0.5	0.5	134.5	83.5	0.5	0.5	92.7	69.6	39.2	0.5	0.5
000037	0.0179	-1.0	-1.0	-1.0	-1.0	-1.0	-1.0	-1.0	-1.0	-1.0	-1.0	-1.0	-1.0	-1.0
00003800	0.0234	64.1	3.5	0.5	0.5	151.4	89.0	0.5	0.5	101.5	101.5	75.4	0.5	0.5
00003900	0.0165	-1.0	-1.0	-1.0	-1.0	-1.0	-1.0	-1.0	-1.0	-1.0	-1.0	-1.0	-1.0	-1.0
00004000	0.0227	113.3	10.1	0.5	0.5	302.1	128.4	0.5	0.5	302.1	128.4	52.6	0.5	0.5
00004100	0.0205	0.5	0.5	0.5	0.5	33.6	97.6	0.5	0.5	107.6	84.1	41.8	0.5	0.5
00004200	0.0191	0.5	0.5	0.5	0.5	20.1	80.5	0.5	0.5	36.2	60.4	48.3	0.5	0.5
00004300	0.0220	0.5	0.5	0.5	0.5	42.4	78.8	0.5	0.5	0.5	42.4	75.7	0.5	0.5
00004400	0.0156	0.5	0.5	0.5	0.5	21.4	135.3	0.5	0.5	0.5	21.4	142.7	0.5	0.5
00004500	0.0124	0.5	0.5	0.5	0.5	68.2	74.4	0.5	0.5	31.0	31.0	49.2	0.5	0.5
00004600	0.0158	0.5	0.5	0.5	0.5	226.9	205.1	0.5	0.5	196.4	144.0	42.9	0.5	0.5
00004700	0.0172	-1.0	-1.0	-1.0	-1.0	-1.0	-1.0	-1.0	-1.0	-1.0	-1.0	-1.0	-1.0	-1.0
00004800	0.0161	0.5	0.5	0.5	0.5	23.0	188.6	0.5	0.5	115.0	110.4	43.6	0.5	0.5
00004900	0.0143	0.5	0.5	0.5	0.5	124.8	204.8	0.5	0.5	129.9	112.4	104.2	0.5	0.5
00005000	0.0205	0.5	0.5	0.5	0.5	39.4	136.6	0.5	0.5	117.1	101.3	33.7	0.5	0.5
00005100	0.9800	0.5	0.5	0.5	0.5	0.5	0.5	0.5	0.5	0.5	0.5	0.5	10.6	12.6
00005200	0.440	0.5	0.5	0.5	0.5	0.5	0.5	0.5	0.5	0.5	0.5	0.5	14.8	15.9
00005300	1.3300	0.5	0.5	0.5	0.5	0.5	0.5	0.5	0.5	0.5	0.5	0.5	8.5	11.6
00005400	1.0400	0.5	0.5	0.5	0.5	0.5	0.5	0.5	0.5	0.5	0.5	0.5	8.6	11.5
00005500	0.8800	0.5	0.5	0.5	0.5	0.5	0.5	0.5	0.5	0.5	0.5	0.5	9.4	11.9
00005600	0.5800	0.5	0.5	0.5	0.5	0.5	0.5	0.5	0.5	0.5	0.5	0.5	15.8	9.4
00005700	1.1100	0.5	0.5	0.5	0.5	0.5	0.5	0.5	0.5	0.5	0.5	0.5	15.8	11.3
00005800	0.6200	0.5	0.5	0.5	0.5	0.5	0.5	0.5	0.5	0.5	0.5	0.5	13.1	15.4
00005900	0.6300	0.5	0.5	0.5	0.5	0.5	0.5	0.5	0.5	0.5	0.5	0.5	14.1	14.9
00006000	0.4400	0.5	0.5	0.5	0.5	0.5	0.5	0.5	0.5	0.5	0.5	0.5	14.6	10.1
00006100	3.3600	0.5	0.5	0.5	4.0	0.5	0.5	0.5	0.5	0.5	0.5	0.5	36.7	52.6
00006200	3.4600	0.5	0.5	0.5	6.5	0.5	0.5	0.5	0.5	0.5	0.5	0.5	27.2	42.6
00006300	2.6800	0.5	0.5	0.5	8.5	0.5	0.5	0.5	0.5	0.5	0.5	0.5	36.2	21.9
00006400	2.6600	0.5	0.5	0.5	8.6	0.5	0.5	0.5	0.5	0.5	0.5	0.5	36.6	56.7
00006500	2.3200	0.5	0.5	0.5	7.6	0.5	0.5	0.5	0.5	0.5	0.5	0.5	38.4	47.9
00006600	2.2000	0.5	0.5	0.5	6.9	0.5	0.5	0.5	0.5	0.5	0.5	0.5	15.0	38.6
00006700	1.2400	0.5	0.5	0.5	16.2	0.5	0.5	0.5	0.5	0.5	0.5	0.5	9.0	14.4
00006800	1.0200	0.5	0.5	0.5	5.3	0.5	0.5	0.5	0.5	0.5	0.5	0.5	26.3	29.6
00006900	2.9200	0.5	0.5	0.5	6.2	0.5	0.5	0.5	0.5	0.5	0.5	0.5	12.7	41.6
00007000	2.9200	0.5	0.5	0.5	8.3	0.5	0.5	0.5	0.5	0.5	0.5	0.5	34.5	54.8

Table D.1 continued

Code No.	Weight (g)	Concentration (μg/g)													
00007100	4.8000	0.5	0.5	0.5	0.5	15.6	0.5	1.4	20.0	4.4	2.3	0.5	1.3	38.3	
00007200	14.4000	0.5	0.5	0.5	0.5	6.7	0.5	2.6	11.8	1.1	1.2	0.5	6.7	40.7	
00007300	20.8000	0.5	0.5	0.5	0.5	4.8	0.5	5.7	11.3	2.7	1.7	0.5	11.9	45.3	
00007400	12.8000	0.5	0.5	0.5	0.5	6.7	0.5	5.5	8.1	5.5	2.5	0.5	7.8	40.7	
00007500	9.2000	0.5	0.5	0.5	0.5	5.2	0.5	3.4	7.5	3.5	2.1	0.5	4.6	38.8	4 weeks
00007600	14.0000	0.5	0.5	0.5	0.5	6.6	0.5	2.2	5.5	3.9	1.4	0.5	9.2	44.2	
00007700	9.6000	0.5	0.5	0.5	0.5	5.4	0.5	7.2	10.8	5.4	1.6	0.5	9.0	40.5	
00007800	8.8000	0.5	0.5	0.5	0.5	3.5	0.5	2.1	4.0	3.6	2.7	0.5	12.1	45.4	
00007900	6.4000	0.5	0.5	0.5	0.5	6.1	0.5	6.1	12.2	3.1	1.9	0.5	2.8	44.4	
00008000	11.2000	0.5	0.5	0.5	0.5	5.2	0.5	2.4	28.9	2.4	3.1	0.5	4.0	39.7	—
00008100	34.7000	0.5	0.5	0.5	1.5	4.5	0.5	1.2	22.1	2.5	0.5	0.5	1.2	46.9	
00008200	22.1000	0.5	0.5	0.5	0.5	1.3	0.5	3.4	5.1	0.5	0.5	0.5	2.3	42.5	
00008300	29.0000	0.5	0.5	0.5	2.2	5.2	0.5	2.1	20.5	0.5	1.4	0.5	1.5	42.1	
00008400	33.9000	0.5	0.5	0.5	0.5	0.5	0.5	2.6	6.0	0.5	1.3	0.5	3.0	44.1	
00008500	30.4000	0.6	0.5	0.5	3.9	1.6	0.5	1.9	4.0	0.5	2.4	0.5	1.2	44.7	5 weeks
00008600	34.4000	0.5	0.5	0.5	1.5	6.0	0.5	1.4	24.6	3.2	3.3	0.5	2.9	43.9	
00008700	29.6000	0.5	0.5	0.5	0.5	0.5	0.5	2.4	5.4	0.5	3.1	0.5	1.8	45.3	
00008800	19.7000	0.5	0.5	0.5	4.8	1.8	0.5	2.2	10.1	0.5	0.6	0.5	1.9	43.0	
00008900	28.0000	0.5	0.5	0.5	0.5	3.8	0.5	1.8	7.0	3.7	0.5	0.5	3.2	42.7	
00009000	26.8000	-1.0	-1.0	-1.0	-1.0	-1.0	-1.0	-1.0	-1.0	-1.0	-1.0	-1.0	-1.0	-1.0	—
00009100	43.4000	0.5	0.5	0.5	9.9	3.9	0.5	19.8	1.7	1.6	0.5	0.5	7.6	43.0	
00009200	41.3000	0.5	0.5	0.5	4.8	2.9	0.5	6.9	1.9	1.8	0.5	0.5	9.7	41.1	
00009300	33.6000	0.5	0.5	0.5	6.7	2.8	0.5	3.3	2.2	3.9	2.1	0.5	4.5	35.9	
00009400	42.7000	0.5	0.5	0.5	7.1	5.1	0.5	12.6	3.6	2.6	0.9	0.5	4.0	38.8	
00009500	40.7000	0.5	0.5	0.5	4.5	4.0	0.5	0.5	4.0	1.5	1.7	0.5	5.0	41.9	
00009600	44.9000	0.5	0.5	0.5	2.2	0.5	0.5	8.3	7.6	3.0	1.4	0.5	2.2	42.4	6 weeks
00009700	45.5000	0.5	0.5	0.5	9.4	1.6	0.5	8.2	2.5	2.5	2.5	0.5	4.4	26.7	
00009800	49.0000	0.5	0.5	0.5	5.3	2.6	0.5	6.1	5.7	1.8	1.3	0.5	11.7	26.4	
00009900	41.8000	0.5	0.5	0.5	6.1	9.1	0.5	6.1	12.6	2.0	0.5	0.5	0.5	45.4	
00010000	46.1000	0.5	0.5	0.5	10.9	3.3	0.5	11.7	3.3	1.0	1.4	0.5	3.7	41.0	
00010100	114.0000	0.5	0.5	0.5	6.8	3.2	0.5	4.5	9.1	2.1	2.7	0.5	22.1	20.6	
00010200	69.0000	0.5	0.5	0.5	4.1	3.8	0.5	3.7	9.0	2.6	2.9	0.5	20.0	26.9	
00010300	38.0000	0.5	0.5	0.5	3.3	4.4	0.5	5.2	14.3	1.9	0.5	0.5	18.2	17.8	—

Label																Note
00010400	23.0000	0.5	0.5	0.5	0.5	2.0	4.1	2.7	1.5	3.5	1.7	2.8	0.5	29.0	24.8	
00010500	30.0000	0.5	0.5	0.5	0.5	2.1	2.6	2.7	1.5	3.5	1.7	2.8	0.5	20.5	21.6	7 weeks
00010600	116.0000	0.5	0.5	0.5	0.5	3.0	2.4	4.7	4.3	3.7	1.5	0.9	0.5	14.8	19.7	
00010700	79.0000	0.5	0.5	0.5	0.5	4.5	2.5	5.0	2.7	1.6	2.4	2.6	0.5	23.8	16.1	
00010800	47.0000	0.5	0.5	0.5	0.5	1.8	2.4	3.1	3.1	6.3	2.3	1.9	0.5	11.7	16.2	
00010900	83.0000	0.5	0.5	0.5	0.5	2.6	2.7	3.7	4.7	7.8	2.1	2.1	0.5	14.1	18.3	
00011000	119.0000	0.5	0.5	0.5	0.5	2.6	2.8	3.2	3.1	7.1	2.5	2.7	0.5	21.1	24.2	
00011100	167.0000	0.5	0.5	0.5	0.5	9.1	0.5	4.2	7.9	5.5	2.7	1.7	0.5	18.9	18.4	
00011200	223.0000	0.5	0.5	0.5	0.5	7.4	1.6	1.0	7.4	2.8	1.3	2.7	0.5	14.9	19.3	
00011300	103.0000	0.5	0.5	0.5	0.5	0.5	1.0	1.5	0.8	0.8	2.9	2.8	0.5	6.1	22.7	
00011400	130.0000	0.5	0.5	0.5	0.5	1.1	0.7	1.8	1.4	1.1	2.5	1.4	0.5	22.7	21.8	
00011500	194.0000	0.5	0.5	0.5	0.5	1.1	1.2	1.2	1.6	1.0	1.9	1.9	0.5	13.7	16.1	8 weeks
00011600	150.0000	0.5	0.5	0.5	0.5	1.0	0.5	1.2	1.8	1.2	1.9	3.1	0.5	14.9	17.0	
00011700	142.0000	0.5	0.5	0.5	0.5	2.0	0.5	3.7	1.4	1.8	1.8	1.0	0.5	9.4	19.9	
00011800	296.0000	0.5	0.5	0.5	0.5	2.8	1.4	0.5	3.5	1.9	2.5	2.5	0.5	5.1	23.2	
00011900	190.0000	0.5	0.5	0.5	0.5	1.9	1.4	0.5	0.7	1.0	2.4	1.8	0.5	15.2	21.0	
00012000	126.0000	0.5	0.5	0.5	0.5	2.6	1.2	1.4	3.4	1.2	2.6	2.0	0.5	17.7	19.9	
00012100	0.0027	156.4	0.5	0.5	41.2	1357.0	541.0	0.5	0.5	0.5	371.0	337.0	0.5	0.5	0.5	
00012200	0.0031	281.9	0.5	0.5	40.4	1264.0	580.0	0.5	0.5	0.5	484.0	794.0	0.5	0.5	0.5	
00012300	0.0034	221.7	0.5	0.5	27.2	1159.0	494.0	0.5	0.5	0.5	721.0	681.0	0.5	0.5	0.5	
00012400	0.0025	172.4	0.5	0.5	28.5	1621.0	672.0	0.5	0.5	0.5	544.0	443.0	0.5	0.5	0.5	
00012500	0.0031	183.3	0.5	0.5	30.1	1342.0	840.0	0.5	0.5	0.5	604.0	605.0	0.5	0.5	0.5	
00012600	0.0031	206.9	0.5	0.5	59.6	1297.0	494.0	0.5	0.5	0.5	721.0	397.0	0.5	0.5	0.5	Fresh seed
00012700	0.0027	210.5	0.5	0.5	54.2	1071.0	321.0	0.5	0.5	0.5	550.0	458.0	0.5	0.5	0.5	
00012800	0.0032	198.7	0.5	0.5	56.7	1281.0	547.0	0.5	0.5	0.5	449.0	479.0	0.5	0.5	0.5	
00012900	0.0029	199.8	0.5	0.5	60.1	1041.0	589.0	0.5	0.5	0.5	437.0	503.0	0.5	0.5	0.5	
00013000	0.0029	211.6	0.5	0.5	47.2	1122.0	348.0	0.5	0.5	0.5	748.0	554.0	0.5	0.5	0.5	
00013100	'EOD'															

Exercise D.1 Chemical structure of plants

The data given in Table D.1 are the weights in grams of fodder rape seeds and plants at 13 growth stages, together with the concentration (μg/g) of 14 volatile chemicals extracted from them. As the data are presented, each line represents a plant, the first value being weight. There are 10 plants of each age and the ages are: stored seed, 1, 2, 3, 4 days, 2, 3, 4, 5, 6, 7, 8 weeks, fresh seed. A value of 0.5 for a concentration means that the chemical in question was not detected, while a value of -1.0 indicates a missing value. (Note that Table D.1 is exactly how the data were presented to me, except for the two lines near the bottom of the table and the encircled observations at week 7 which are referred to in the notes overleaf.)

The purpose of the exercise as presented to me was to 'use an appropriate form of multivariate analyis to distinguish between the ages of the plants in terms of the amounts of different chemicals which are present'. In other words, we want a (preferably simple) description of the chemical structure of the plants and seeds at each stage so that the different stages can easily be distinguished and so that the ageing process can be described in chemical terms.

In practice the exercise would best be carried out jointly by a biochemist and a statistician and the results related to other studies on the attraction of insects to host plants. However, in this exercise the reader should simply try to summarize the data in whatever way seems appropriate.

Exercise D.2 Backache in pregnancy

Backache is a frequent complaint among pregnant women. An investigation (Mantle, Greenwood and Currey, 1977) was carried out on all the 180 women giving birth in the labour wards of the London Hospital during the period May to August 1973. Each woman received the questionnaire within 24 hours of delivery and had the assistance of a physiotherapist in interpreting and answering the questions. There was a 100% response rate and 33 items of information were collected from each patient. The data are shown in Table D.3 together with a detailed list of the items recorded and the format. As well as a (subjective) measure of back pain severity, each patient's record contains various personal attributes, such as height and weight, and whether or not a list of factors relieved or aggravated the backache.

The object of the analysis is to summarize the data and, in particular, to see if there is any association between the severity of backache and other variables, such as number of previous children and age. As part of this exercise you should scrutinize the data in order to assess their quality and reliability. In practice this analysis should be carried out with an appropriate medical specialist, but here you are simply expected to use your common sense. Generally speaking, only simple exploratory techniques should be used to highlight the more obvious features of the data.

(*Notes for lecturers*: This exercise is intended to be carried out on a computer by students who have access to the data on file. Working 'by hand' would be too time-consuming and punching the data would also take too long. Teachers may obtain a free copy of the data from the author by computer e-mail as an ASCII file. This is well worth doing as the data are ideal for analysis by students at all levels. The above project to 'summarize the data' is really for third-level students. First-level students will find the data suitable for trying out descriptive statistics. I have also used the data for second-level students with much more specific objectives such as:

1. Comment briefly on the data-collection method.
2. Comment briefly on the quality of the data, and
   (a) pick out one observation which you think is an error.
   (b) pick out one observation which you think is an outlier.
3. Construct
   (a) a histogram, and
   (b) a stem-and-leaf plot
   of the gain in weight of the patient during pregnancy. Which of the graphs do you prefer?
4. Construct an appropriate graph or table to show the relationship, if any, between backache severity (item 2) and
   (a) age (item 4),
   (b) number of previous children (item 9).
5. Briefly say which factors, if any, you think may be related to back pain severity.
6. Construct a table showing the frequencies with which items 11–18 relieve backache.

Indeed, it is easy to think of a large number of different exercises, so I have found the data ideal for use in different projects with different students.)

Notes on Exercise D.1

Have you read the comments at the beginning of this section? Then you know that the first step is to scrutinize the data, assess their structure, pick out any obvious errors and outliers, and also notice any obvious properties of the data.

You may think the data look a 'bit of a mess' and you would be right. You have to get used to the look of real data, which generally need to be 'cleaned up' prior to analysis. For example, the code numbers could (and should) be given as simple integers without all the zeros. Also notice the large number of undetected values recorded as 0.5 which could (and should) be replaced by zeros or blanks.

With the data stored in a computer, it is relatively easy to print revised data to replace the original, rather 'horrible' data format. The observations could

Table D.3 Data on backache in pregnancy compiled from 180 questionnaires
Key:

Item	Description	Fortran format
1	Patient's number	I3
2	Back pain severity: 'nil' – 0; 'nothing worth troubling about' – 1; 'troublesome but not severe' – 2; 'severe' – 3	I2
3	Month of pregnancy pain started	I2
4	Age of patient in years	I3
5	Height of patient in metres	F5.2
6	Weight of patient at start of pregnancy in kilograms	F5.1
7	Weight at end of pregnancy	F5.1
8	Weight of baby in kilograms	F5.2
9	Number of children from previous pregnancies	I2
10	Did patient have backache with previous pregnancy: not applicable – 1; no – 2; yes, mild – 3; yes, severe – 4	I1
11–18	Factors relieving backache: no – 0; yes – 1	
	1. Tablets, e.g. aspirin	I1
	2. Hot water bottle	I1
	3. Hot bath	I1
	4. Cushion behind back in chair	I1
	5. Standing	I1
	6. Sitting	I1
	7. Lying	I1
	8. Walking	I1
19–33	Factors aggravating pain: no – 0; yes – 1	
	1. Fatigue	I1
	2. Bending	I1
	3. Lifting	I1
	4. Making beds	I1
	5. Washing up	I1
	6. Ironing	I1
	7. A bowel action	I1
	8. Intercourse	I1
	9. Coughing	I1
	10. Sneezing	I1
	11. Turning in bed	I1
	12. Standing	I1
	13. Sitting	I1
	14. Lying	I1
	15. Walking	I1

Table D.3 For key see facing page.

ID								
001	1	0	26	1.52	54.5	75.0	3.35	0100000000000000000000000
002	3	0	23	1.60	59.1	68.6	2.22	1210010000000100100000000
003	2	6	24	1.57	73.2	82.7	4.15	0100011010100101000000000
004	1	8	22	1.52	41.4	47.3	2.81	0100000000000001000000000
005	1	0	27	1.60	55.5	60.0	3.75	0100000000000000000000000
006	1	7	32	1.75	70.5	85.5	4.01	1200000000000000000000010
007	1	0	24	1.73	76.4	89.1	3.41	0100000000000000000000000
008	1	8	25	1.63	70.0	85.0	4.01	1100010001000000000000000
009	2	6	20	1.55	52.3	59.5	3.69	1400000100000000001000000
010	2	8	18	1.63	83.2	90.9	3.30	0110000100000100000000010
011	1	0	21	1.65	64.5	75.5	2.95	0100000000000000000000000
012	1	0	26	1.55	49.5	53.6	2.64	0100000010000000000000000
013	2	6	35	1.65	70.0	82.7	3.64	7200100000100000000000100
014	1	8	26	1.60	52.3	64.5	4.49	1300110001000000000000100
015	1	6	34	1.68	68.2	77.3	3.75	3300100101000001000001000
016	0	0	25	1.50	47.3	55.0	2.73	1000000000000000000000000
017	1	7	42	1.52	66.8	73.2	2.44	6400010000010000000000000
018	2	6	26	1.65	70.0	81.4	3.01	1300010000010101000000010
019	2	6	18	1.60	56.4	70.0	3.89	1300010100000100001001001
020	0	1	42	1.65	53.6	63.6	2.73	2000000000000000000000000
021	2	0	28	1.63	59.1	72.3	3.75	1300000000000110000001001
022	1	0	26	1.52	44.5	56.4	3.49	0101000000010000000000000
023	1	0	23	1.57	55.9	60.9	3.07	0100000000000000000000000
024	2	6	21	1.55	57.3	77.3	3.35	0100010000000000000001001
025	2	7	32	1.52	69.5	75.5	3.64	5300100010101000000000000
026	1	8	18	1.60	73.2	81.4	2.05	0100010000010001000010000
027	1	0	25	1.70	52.3	59.5	2.44	1310000000010101010010000
028	1	0	30	1.63	62.7	72.3	3.07	4000000000000000000000000
029	1	0	19	1.65	73.6	92.7	3.35	0000000000000000000000000
030	2	7	26	1.65	70.0	89.1	3.21	1200000000111111000000000
031	1	8	28	1.68	56.8	70.9	3.41	0100000000000000000001000
032	1	0	21	1.60	58.2	69.5	3.30	0000000000011111000000101
033	0	0	29	1.57	68.2	75.0	3.35	0000000000000000000000000
034	1	8	27	1.65	50.9	66.4	3.10	1200010000000000000000100
035	2	2	30	1.60	50.9	62.7	3.75	1210000000000000000001000
036	2	5	26	1.75	69.5	84.1	4.20	0100010000000000000001000
037	1	8	21	1.60	62.7	73.2	2.95	0100100000000000000000001
038	2	2	24	1.73	63.6	69.5	3.18	1200000100111001000001000
039	2	8	28	1.50	55.5	66.4	4.35	2400000100000000000000000
040	1	8	27	1.57	55.9	62.3	3.69	0101100000101000000000000
041	3	6	32	1.60	47.7	66.8	3.41	5310100000100100001000001
042	3	5	26	1.63	55.5	63.2	3.30	0100011001100101000000110
043	2	7	37	1.63	60.9	69.5	3.41	2200010101000000000000100
044	3	7	31	1.60	60.5	77.3	3.27	1000100100000000000000000
045	1	0	24	1.57	62.3	77.3	3.38	0000001010000100000000000
046	3	5	24	1.65	57.3	73.6	3.69	0000010100011010000001001
047	1	0	23	1.52	55.5	69.5	3.35	0100000000000000000000000
048	2	3	31	1.60	47.7	55.9	1.62	0110000010101000000000000
049	1	8	37	1.60	76.4	89.1	3.58	1300010000010100000000000
050	2	6	23	1.65	95.5	95.5	2.73	1200100100000000000001011
051	1	4	39	1.70	72.3	89.1	3.44	0000100010000100000000100
052	2	6	32	1.70	65.9	75.0	3.66	1400000100010000000001110
053	2	4	30	1.52	50.9	57.3	2.56	1301101000101111000001000
054	2	3	24	1.57	57.3	58.6	1.08	0100010100000000000000000
055	1	0	24	1.75	69.1	77.3	3.38	0100000010000000000001000
056	2	6	29	1.55	47.7	54.5	3.81	2300000100000000000001000
057	1	0	24	1.75	59.1	68.6	2.98	3200000000000000000000000
058	2	7	27	1.70	60.0	71.4	2.95	1210010010010001100100100
059	2	4	30	1.57	60.0	75.9	2.84	1201000100000000001000000
060	2	3	32	1.63	66.4	82.7	2.64	2310011110001000000001110

Table D.3 continued

061	1	0	23	1.55	44.5	48.2	2.27	0100000000000000000000000
062	1	0	18	1.60	63.6	75.0	3.47	0100000000000000000000000
063	2	9	28	1.55	47.7	58.2	2.56	1200010000100001000001000
064	1	0	24	1.50	45.9	55.0	3.10	1400000010001000000000000
065	1	5	22	1.68	53.2	64.5	3.07	0100000100011000000000001
066	2	5	29	1.57	60.5	74.1	3.27	1300010100000010000000000
067	3	2	30	1.65	89.5	97.7	3.81	1301000001000100000000000
068	2	4	22	1.70	61.8	75.5	3.21	0111000010111100000001000
069	2	5	31	1.63	66.4	82.3	2.27	0100010000000001000000000
070	1	0	29	1.57	60.5	64.1	3.30	3200000000000000000000000
071	1	0	28	1.65	60.5	73.2	3.75	0100000000000000000000000
072	3	6	32	1.60	47.7	50.5	1.90	3300110000100000000001100
073	1	7	21	1.55	67.3	80.0	3.27	0100000001000000000000000
074	2	5	24	1.50	50.0	63.6	2.76	0101000010000000000000000
075	2	5	26	1.65	55.5	67.3	4.15	2400010100101011000001000
076	1	0	26	1.73	57.3	65.5	3.07	1200000000010000000000000
077	2	7	26	1.60	65.0	80.5	3.95	0100010000000100000000000
078	3	0	28	1.65	57.3	82.7	3.89	0100100101100110000001100
079	2	0	36	1.73	75.9	87.3	3.64	1300100101111110000000001
080	2	7	29	1.55	75.5	78.6	2.61	0100000010100000000010000
081	1	7	23	1.65	47.7	51.8	2.84	1300010000010100000001001
082	3	8	19	1.70	57.3	74.5	3.69	2201110100010100000000001
083	3	7	30	1.73	63.6	80.5	3.61	0410010100000000000001000
084	2	6	24	1.60	50.0	59.1	3.52	1200110100111100000000000
085	1	0	24	1.70	60.0	65.0	3.18	0100000001000000000000000
086	1	0	30	1.55	47.7	59.1	3.15	1200010000010000000000000
087	2	5	25	1.65	50.5	60.9	3.35	1200010000010000000000000
088	3	7	23	1.55	62.7	85.5	4.18	0100010000111110000001001
089	3	3	33	1.63	63.6	70.9	2.93	1310010010100001000001000
090	1	0	31	1.52	52.3	53.2	2.39	2200000000000000000000000
091	2	7	21	1.50	63.6	63.6	2.98	2200000000001000000000000
092	1	0	27	1.75	68.2	82.7	5.97	0000000000100000000000000
093	2	5	25	1.60	68.2	76.4	3.18	0100010000000000000000100
094	3	2	34	1.63	65.9	80.5	4.15	3210000000000000000000101
095	2	6	26	1.70	65.9	68.2	3.35	1400000010000000000000000
096	2	0	20	1.57	60.5	69.5	3.24	0000010000000000000000000
097	2	0	28	1.68	89.1	94.5	2.59	0000100000010100000000000
098	2	0	26	1.70	80.5	89.1	1.42	0010010010111110000001001
099	2	6	22	1.60	63.6	82.7	3.89	2300010000000000000001001
100	2	4	26	1.50	43.6	53.6	3.81	0100000100000000000000101
101	1	8	21	1.55	54.5	67.3	3.72	0100010100000000000000100
102	1	7	18	1.60	62.7	71.4	2.44	1000000001000000000000000
103	2	6	22	1.52	54.1	65.0	2.78	0101000100000000000001010
104	0	0	20	1.63	53.6	73.2	2.78	0000000100000000000000000
105	1	0	25	1.65	61.8	66.8	3.18	0100000000010000000000000
106	2	4	23	1.57	76.4	87.3	2.78	1100100000010000000000100
107	1	5	22	1.60	74.5	82.7	2.90	0100000010000000000010000
108	2	6	22	1.68	47.7	53.6	3.01	3300100001000000000001000
109	1	8	32	1.63	52.3	56.8	2.50	0100000101000000000000000
110	2	6	26	1.57	53.6	65.0	2.87	1200010000000000010000000
111	1	0	25	1.63	58.6	62.3	2.44	0100000000000000000000000
112	0	1	19	1.57	51.8	71.8	3.35	0000000000000000000000000
113	1	0	21	1.60	76.4	78.2	3.07	0000000000000000000000000
114	1	0	23	1.60	56.4	71.4	2.50	0100000000000000000000000
115	2	7	20	1.60	54.1	68.2	3.58	0100010001010000000000101
116	1	4	19	1.65	57.3	70.0	2.61	0100010010001000000000000
117	2	8	30	1.50	53.2	66.4	3.98	1100100000010000000001000
118	1	0	26	1.60	57.3	65.9	3.89	0000000000000000000000010
119	1	0	24	1.57	60.5	71.5	3.24	0100000101000000000001000
120	1	6	28	1.60	46.8	60.5	2.98	1300000101000000000000001

Table D.3 continued

121	0	0	16	1.63	60.5	65.9	3.18	0000000000000000000000000
122	1	0	29	1.63	54.5	67.3	3.58	0100000000000000000000000
123	3	6	25	1.57	59.1	69.5	3.89	1200000010101000000011001
124	2	3	17	1.52	48.2	61.8	2.78	0000100100010000111000000
125	1	7	23	1.65	77.3	83.2	3.58	0100000010000000000000100
126	1	0	26	1.63	72.7	80.0	2.95	0100000000000000000000000
127	1	0	20	1.63	58.6	70.0	2.84	1100000100000000000000000
128	2	2	25	1.68	54.5	71.8	3.72	0100010010011010000000000
129	1	3	24	1.63	61.8	73.6	2.84	0100000010000000000001110
130	3	3	35	1.63	69.5	72.3	3.52	2200000010000100000000000
131	0	0	28	1.70	78.2	98.6	2.84	1000000000010000000000010
132	3	4	25	1.63	79.5	90.5	3.15	0100000000000000000000010
133	1	6	22	1.63	60.5	77.7	2.39	0111000000000000000001000
134	1	7	42	1.50	44.5	48.6	2.27	5210000010000000000001001
135	1	8	21	1.60	60.5	73.2	3.27	0100000010000000000000100
136	2	4	20	1.52	60.5	75.0	3.64	0000010000000010010000000
137	1	0	24	1.63	56.4	70.0	3.07	0000010000000000000000000
138	1	5	33	1.65	40.0	55.0	3.47	0110000010010000101000000
139	1	6	33	1.65	63.6	82.7	3.86	0100100010100000000000000
140	3	3	29	1.63	58.2	65.5	2.84	1310000000000000000000000
141	2	8	27	1.65	57.3	69.5	3.64	0100100101101000000001000
142	3	3	20	1.52	44.5	51.4	1.93	1300000001001010101011111
143	2	6	18	1.63	49.1	56.8	3.24	0000000000100000010100100
144	2	7	26	1.57	52.7	65.9	3.49	0100010101000010000000000
145	1	7	21	1.55	52.3	65.0	3.47	0100000010100000001000000
146	2	4	21	1.52	50.9	68.2	3.64	0100000010001010000001000
147	1	5	21	1.68	66.8	87.3	2.90	0100000110011101000000000
148	1	0	26	1.65	50.9	70.0	3.52	0200000000000000000000001
149	1	8	21	1.55	53.2	56.4	2.44	0100000010000000000000000
150	2	7	22	1.47	44.5	58.2	2.87	0100010000000101000010000
151	3	7	39	1.50	50.9	72.7	3.04	4200001000000000001001
152	2	6	27	1.63	60.5	71.4	3.21	0100001000000000001011000
153	1	0	24	1.57	60.0	72.7	3.35	0100000010101000000000000
154	1	4	24	1.63	64.5	75.0	2.87	0110000010000000000000100
155	1	7	19	1.55	52.7	57.7	2.61	0100010000000000000000010
156	1	0	22	1.60	67.3	73.6	2.05	0000000000000000000000000
157	3	8	27	1.55	60.5	73.2	3.98	0010100000010000000000100
158	3	2	35	1.73	72.3	89.1	6.28	3310010000010000000000100
159	0	0	24	1.65	83.6	86.8	3.64	0000000010010000000000000
160	1	0	35	1.55	52.3	59.1	3.64	0100000010100000000000000
161	1	7	21	1.52	59.1	67.3	2.95	0100000101111010000001001
162	3	7	21	1.68	54.1	74.5	3.72	1200000010111101000000000
163	3	9	38	1.52	43.6	60.9	3.24	1410000000000000000001000
164	1	0	32	1.73	54.5	62.7	3.41	1300010010101000000001000
165	3	8	15	1.55	48.2	56.4	2.70	0110000100000000000000000
166	2	6	23	1.57	60.5	100.0	3.58	0100010000000000000000100
167	2	7	19	1.50	49.5	63.6	3.35	0000010100000000000001101
168	2	8	25	1.60	59.5	72.3	2.73	1300000101010000000000000
169	2	5	22	1.50	53.6	63.2	3.35	0100011000000000000000100
170	0	0	19	1.55	57.3	65.0	3.07	1200000000000000000000000
171	1	0	23	1.68	57.3	61.4	3.58	0100000000000000000001010
172	3	0	36	1.63	54.1	60.0	2.84	0101000001000000000001010
173	0	0	21	1.55	50.9	63.6	3.01	0100000001000000000000100
174	0	0	30	1.52	38.2	48.2	2.56	0000000100000000000000000
175	0	0	42	1.70	55.0	65.5	3.27	4200000000000000000000000
176	1	0	34	1.63	50.9	60.5	2.93	0100001000000000000000100
177	3	3	26	1.63	66.8	84.1	3.10	1300100010000000000001100
178	1	7	18	1.50	54.1	60.5	3.52	1300100001000000000000100
179	3	6	39	1.52	82.7	84.1	3.35	1210000000000000000000000
180	1	0	25	1.52	52.3	66.8	3.24	1200010000000010000010000

be rounded in an appropriate way, usually to two-variable-digit accuracy, the missing values could be replaced by asterisks or removed completely, and undetected values removed as noted above. The values for fresh seed should be reordered at the beginning of the data as they need to be compared with the stored-seed values rather than with the 8-week data. Inspection of the data can also be made easier by drawing lines on the data table to separate the different growth stages (as at the bottom of Table D.1), and by labelling the 14 variables from A to N.

The data should then be inspected for suspect values which could be encircled. There are several cases where the same figure is repeated in two or more columns (e.g. the values at 7 weeks encircled in Table D.1) where digits are possibly transposed, or where there are possible outliers. With no access to the original data, it is a matter of judgement as to which observations should be adjusted or disregarded.

Summarizing the data

It may seem natural to begin by calculating the sample mean for each variable at each growth stage. However, several variables appear to have a skewed distribution so that medians may be more appropriate than means. Alternatively, some sort of transformation may be desirable. Instead of analysing concentrations, as in the given data, perhaps we should multiply by the appropriate weight and analyse amounts.

A completely different approach, which turned out to be suitable in this case, is to treat the data as binary (the chemical is observed/not observed). If insects are attracted by certain chemicals and repelled by others, then the presence or absence of certain chemicals may be more important than concentration or amounts. Table D.2 shows the presence/absence table as well as the mean weights at different growth stages. The order of the variables has been changed to emphasize the block structure of the table. (Did you think of doing that?)

Table D.2 gives clear guidance on distinguishing between the different growth stages. For example, chemical C is only present in stored seed and 1-day-old plants. Week 2 is characterized by the presence of M and N and so on. The table also suggests that it is sensible to query the appearance of chemical D in week 3, and reveals that the period of greatest change is between day 4 and 2 weeks. The absence of data in this period is therefore unfortunate.

Discussion

The presence/absence table gives a clear guide to differences between growth stages. It is not obvious that anything further can usefully be done with these data in the absence of biochemical advice. Certainly there seems little point in

Table D.2 Presence/absence table

Growth stage	D	C	B	A	L	K	F	J	E	M	N	H	I	G	Mean weight (g)
Fresh seed	+			+		+	+	+	+						0.0030
Stored seed	+	+	+	+	+	+	+	+	+						0.0034
Day 1		+	+	+	+	+	+	+	+						0.0071
Day 2			+	+	+	+	+	+	+						0.012
Day 3			+	+	+	+	+	+	+						0.025
Day 4					+	+	+	?	+						0.017
Week 2										+	+				0.80
Week 3	+									+	+				2.4
Week 4					+	+	+			+	+	+	+		11.2
Week 5					?	?	?	?		+	+	+	+		29.0
Week 6					?	?	+	+		+	+	+	+		43.0
Week 7					?	+	+	+		+	+	+	+	+	72.0
Week 8					+	?	+	?		+	+	+	+	?	172.0

Key: + The substance is detected in all samples.
? The substance is detected in some samples.

carrying out any multivariate analysis as suggested in the statement of the problem. Indeed principal component analysis gives no further understanding of the data (see Chatfield, 1982, Exercise 5). Although no explicit statistical model has been fitted, we now have a good idea of the main properties of the data and it is not uncommon for binary information to represent most of the available information in quantitative data (e.g. Exercise E.4).

Moral

Simple descriptive statistics is sometimes adequate for apparently complicated data sets.

Brief comments on Exercise D.2

It would take too much space to comment fully on the data. These brief remarks concentrate on important or unusual aspects and leave the rest to the reader (see also Chatfield, 1985, Example 3).

You should first have queried the data-collection method, which is not difficult to criticize. The sample is not random and the results will not necessarily generalize to women in other parts of Britain or women who were pregnant at a different time of year. It is also debatable whether women should be questioned within 24 hours of delivery, although the 'captive' audience did at least produce a 100% response rate! The questionnaire is open to criticism in places and the assessment of pain necessarily involves a difficult subjective judgement. However, the collection methods could only have been improved at much greater cost which would be unwarranted in an exploratory survey of this nature. While any model-based inference is probably unjustifiable, it would be over-reacting to reject the data completely. As a compromise, they should be treated in a descriptive way to assess which variables are potentially important.

The data should be scrutinized and summarized using an appropriate computer package. I used MINITAB which is very good for this sort of exercise. The data are fairly 'messy' and you should begin by plotting a histogram of each variable, with the particular aim of finding suspect values and other oddities. Four examples must suffice here. The histogram of patients' heights reveals several cells with zero frequencies which destroy the expected 'normal' shape. A little research reveals that heights must have been recorded in inches and then converted to centimetres in order to produce the given data. Did you spot this? The histogram of baby weights reveals two babies whose weights are 5.97 kg and 6.28 kg. These values are much higher than the remaining weights and I would judge them to be outliers. While they may have been misrecorded (or indicate twins?), they are medically feasible and should probably not be excluded. Patient weight gains during pregnancy are worth looking at, by subtracting 'weight at start' from 'weight at end'. One patient has zero weight gain, indicating a possible repetition error. However, the histogram of weight gains indicates that a zero value is not out of line with other values and so need not be regarded as suspicious. Finally, we note that 26 patients recorded a zero value when asked to assess backache in previous pregnancies. A zero value is impossible according to the listed code and so it is probably wise to ignore item 10.

After screening the data, a variety of summary statistics, graphs and tables should be calculated. The most important statistic is that nearly half the sample (48%) had either troublesome or severe backache, indicating that back pain really is a serious problem. In order to assess the relationship between the discrete variable 'severity of backache' and other variables, I note that scatter diagrams are inappropriate, but rather that a set of boxplots (e.g. Fig. D.1) or a two-way table (e.g. Table D.4) should be formed depending on whether the other variable is continuous (e.g. height) or discrete (e.g. number of previous children). When plotting boxplots with unequal sample sizes, the latter should be clearly recorded, as in Fig. D.1, because the range increases with sample size and is potentially misleading.

Without doing any ANOVA, it should now be obvious that there is little or no association between backache and height, weight, weight gain or weight of baby. In Fig. D.1 we see a slight tendency for older women to have more severe backache, while Table D.4 indicates a clearer association with the number of previous children. The latter relationship is easy to explain in that picking up

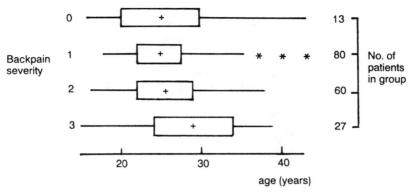

Figure D.1 The set of boxplots showing backpain severity against age.

a young child may injure the back. Of course age and number of previous children are correlated anyway (see Mantle, Greenwood and Currey, 1977, Fig. 1).

To analyse the factors relieving or aggravating backache, the first step is to produce count frequencies for each factor and tabulate them in numerical order, as in Table D.5. A 'cushion behind the back' is best for relieving backache, while 'standing' is the most aggravating factor. The more important factors could, if desired, be tabulated against back pain severity to see which are most associated with severe backache.

While the above analysis looks fairly trivial at first sight, I have found that students often have difficulties or produce inferior results as compared with those produced by a more experienced analyst. For example, did you remember to look for errors and outliers? In Table D.4, did you combine the data for two or more pregnancies (or some similar grouping)? If not, I hope you agree that this modification makes the table easier to 'read'. In Fig. D.1, did you separate the age values into discrete groups and produce a two-way table rather than a set of boxplots? In Table D.5, did you present the factors in the same order

Table D.4 The number of women tabulated by degree of back pain severity and number of previous children

| Number of previous children | Degree of back pain severity | | | | |
	0	1	2	3	Total
0	8(8%)	56(55%)	28(28%)	9(9%)	101(100%)
1	3(6%)	16(31%)	22(42%)	11(21%)	52(100%)
2 or more	2(7%)	8(30%)	10(37%)	7(26%)	27(100%)
Total	13	80	60	27	180

Table D.5 The number of women for whom a particular factor relieved backache

Factor	Frequency
Cushion behind back in chair	57
Lying down	38
Tablets, e.g. aspirin	23
Sitting	20
Hot bath	15
Standing	9
Hot water bottle	9
Walking	5

as listed in Table D.3, and, if so, do you agree that Table D.5 is easier to read? Finally, are all your graphs and tables clearly labelled?

Summary

Backache is troublesome or severe for nearly half the given sample of women. There seems to be little association with other variables except that it is more likely to occur in women with previous children. 'Standing' is the factor which aggravates backache most, while 'a cushion behind the back' is best for relief.

Moral

Elementary 'descriptive' methods are not as easy to handle as is sometimes thought. Students need plenty of practical experience and a 'messy' set of data like the one given here is ideal for learning how to apply graphical and tabular techniques in a more methodical way.

E

Analysing more structured data

Data which arise from experimental designs and more complex surveys often have a well-defined structure which needs to be taken into account during the analysis. For example, in a randomized block design we know that each observation on the response variable corresponds to a particular treatment and to a particular block. Comparisons between treatments should then be made after eliminating variation due to between-block differences.

It will normally be useful to fit a proper stochastic model to structured data. Most such models can be regarded as special cases of the generalized linear model (Appendix A.9), but the simpler formulation of the general (as opposed to generalized) linear model should be used where possible as it is easier to explain to non-statisticians.

Some complicated designs, such as fractional factorial experiments and Latin squares, give rise to highly structured data sets where the form of analysis is largely determined a priori. Then the scope for IDA may be rather limited, apart from data quality checks (e.g. Example E.2). However, in many other cases, IDA has an important role to play, not only for checking data quality and getting a 'feel' for the data, but also in formulating a sensible model based on empirically reasonable assumptions.

The examples in this section concentrate on problems where IDA has an effective role to play.

Exercise E.1 Assessing row and column effects

A textbook reports that the following experimental results (Table E.1) arose when comparing eight treatments ($t_1 - t_8$) in three blocks (I–III).

(a) What sort of design do you think was used here?
(b) From a visual inspection of the data:
   (i) Are the treatment effects significantly different from one another?
   (ii) Are the block effects significantly different from one another?

(c) Do you think this is an artificial data set?

(d) How would you analyse these data 'properly'?

(*Note*: The units of measurement were not stated.)

Table E.1 Some experimental results from treatment comparisons

Block	t_1	t_2	t_3	t_4	t_5	t_6	t_7	t_8	Row mean
				Treatment					
I	35.2	57.4	27.2	20.2	60.2	32.0	36.0	43.8	39.0
II	41.2	63.4	33.2	26.2	60.2	32.0	36.0	43.8	42.0
III	43.6	53.2	29.6	25.6	53.6	32.0	30.0	44.4	39.0
Column mean	40.0	58.0	30.0	24.0	58.0	32.0	34.0	44.0	40.0

Exercise E.2 Engine burners – Latin square design

The data* shown in Table E.3 resulted from an experiment to compare three types of engine burner, denoted by B_1, B_2 and B_3. Tests were made using three engines and were spread over three days and so it was convenient to use a design called a **Latin square** design in which each burner was used once in each engine and once in each day. Analyse the data.

Table E.3 Measurements of efficiency of three burners

	Engine 1	Engine 2	Engine 3
Day 1	B_1: 16	B_2: 17	B_3: 20
Day 2	B_2: 16	B_3: 21	B_1: 15
Day 3	B_3: 15	B_1: 12	B_2: 13

Exercise E.3 Comparing wheat varieties – unbalanced two-way ANOVA

Performance trials were carried out to compare six varieties of wheat at 10 testing centres in three regions of Scotland in 1977. A description of the experiment and the data is given by Patterson and Silvey (1980). It was not possible to test each variety at each centre. The average yields, in tonnes of grain

* Fictitious data based on a real problem.

Table E.6 Yield (in tonnes of grain per hectare) of six varieties of winter wheat at 10 centres

Variety	E1	E2	N3	N4	N5	N6	W7	E8	E9	N10
					Centre					
Huntsman	5.79	6.12	5.12	4.50	5.49	5.86	6.55	7.33	6.37	4.21
Atou	5.96	6.64	4.65	5.07	5.59	6.53	6.91	7.31	6.99	4.62
Armada	5.97	6.92	5.04	4.99	5.59	6.57	7.60	7.75	7.19	—
Mardler	6.56	7.55	5.13	4.60	5.83	6.14	7.91	8.93	8.33	—
Sentry	—	—	—	—	—	—	7.34	8.68	7.91	3.99
Stuart	—	—	—	—	—	—	7.17	8.72	8.04	4.70

per hectare, are given in Table E.6 for each variety at each centre at which it was grown.

(a) Comment on the type of design used here.
(b) Analyse and summarize the data. In particular, if you think there are differences between the varieties, pick out the one which you think gives the highest yield on average.

From other more extensive data, Patterson and Silvey (1980) suggest that the standard error of the average yield for one variety at one centre is known to be around 0.2 (tonnes of grain per hectare), and you may use this information when assessing the data.

Exercise E.4 Failure times – censored data

The data in Table E.9 show the results of an initial testing programme which was designed to investigate the effects of four variables on the failure times of

Table E.9 Failure times (in hours) from a life-testing experiment

Load	Promoter	Pre-charging (hours)	−0.7		−0.9		−1.2		−1.45	
			Charging potential (V)							
			Replication (rig)							
			1	2	1	2	1	2	1	2
400	0	0	NF	NF	NF	NF	4.2	NF	NF	3.6
	0	0.25	NF	NF	NF	3.9	NF	NF	NF	3.6
	5	0	NF	NF	NF	2.0	2.4	1.9	2.2	2.2
	5	0.25	NF	NF	11.3	2.7	1.8	2.0	5.3	1.0

Table E.9 continued

Load	Promoter	Pre-charging (hours)	−0.7		−0.9		−1.2		−1.45	
			Replication (rig)							
			1	2	1	2	1	2	1	2
450	0	0	NF	NF	18.1	NF	7.0	4.9	4.6	3.2
	0	0.25	NF	NF	3.0	9.1	8.2	NF	3.2	2.5
	5	0	NF	NF	9.5	1.8	3.1	1.7	1.7	2.6
	5	0.25	2.9	NF	3.0	1.4	2.0	2.0	1.0	0.7
475	0	0	17.5	8.3	2.5	3.5	NF	1.7	2.7	4.0
	0	0.25	NF	NF	1.5	1.4	4.2	2.0	NF	4.1
	5	0	NF	NF	2.3	2.0	1.9	1.1	1.9	1.3
	5	0.25	NF	3.0	2.0	0.9	1.1	0.5	0.9	0.7
500	0	0	NF	NF	NF	2.3	2.9	2.0	NF	4.5
	0	0.25	NF	NF	NF	2.5	9.2	3.6	1.3	NF
	5	0	NF	NF	2.2	1.7	1.3	1.0	1.0	1.4
	5	0.25	NF	2.8	1.6	1.2	1.5	0.6	1.4	0.7
600	0	0	2.8	1.9	1.8	1.6	1.8	2.2	1.0	1.7
	0	0.25	NF	2.1	3.1	1.2	1.2	1.0	3.7	2.0
	5	0	4.1	2.5	1.2	0.8	0.9	0.9	0.9	0.6
	5	0.25	8.4	2.5	1.0	0.5	2.0	1.9	0.7	0.6
750	0	0	2.0	1.5	1.2	0.6	1.4	0.7	1.6	0.4
	0	0.25	1.0	0.5	1.6	0.6	0.4	0.7	1.3	0.5
	5	0	2.5	1.5	0.9	0.7	0.7	0.2	0.8	0.2
	5	0.25	1.6	1.5	0.9	0.2	0.6	0.0	0.3	0.0

Column group header: Charging potential (V)

NF denotes no failure by 20 hours.

different rigs. The four variables are load, promoter, pre-charging and charging potential. The practical details were clarified at the time the data arose, but do not need to be explained here. However, all tests were truncated at 20 hours and NF is used to indicate that no failure occurred in the first 20 hours.

Identify the design and analyse the data. Which of the variables are important in determining the length of failure time and how can it be maximized?

Exercise E.5 The effect of dust on lung weight

An investigation was carried out into the effect of dust on the lung weights of rats. An increase in lung weight is thought to be positively correlated with the chance of getting tuberculosis. At two laboratories, denoted I and II, 16 rats were randomly divided into two groups denoted A and B. The animals in group

A were kept in an atmosphere containing a known fixed amount of dust, while those in group B were kept in a dust-free atmosphere. After three months, the animals were killed and their lung weights were measured (in grams). The results are given in the following table.

Table E.12 Lung weights (grams)

Lab I		Lab II	
A	B	A	B
5.44	5.12	5.79	4.20
5.35	3.80	5.57	4.08
5.60	4.95	6.52	5.81
6.46	6.43	4.78	3.63
6.75	5.03	5.91	2.80
6.03	5.08	7.02	5.10
4.15	3.22	6.06	3.64
4.44	4.42	6.38	4.53

Analyse the data.

Exercise E.6 Two-level factorial experiments

Two-level factorial experiments are an important type of experimental design, especially in **screening experiments** where the aim is to make a preliminary assessment of several possible explanatory factors which may influence the response variable, possibly in an interactive way. Sometimes fractional factorial experiments are used, but this exercise considers two unreplicated complete factorial experiments of the 2^4 form. Many textbooks describe the standard analysis for such designs which usually relies on assuming that there are no interactions higher than first-order so that the degrees of freedom associated with high-order interactions can be used to estimate the residual variation. The results of such an analysis often comprise a list of estimates of main effects and first-order interactions together with a large ANOVA table which may, or may not, be helpful.

This exercise presents two chemometric examples from Sundberg (1994) who gives the technical background necessary for a full appreciation of the problems. The latter will not be repeated here (though in general it is always wise to get relevant background information) as it is not necessary for the problems as stated below.

(a) The standard analysis of a 2^n complete factorial experiment assumes the response variable is quantitative. Suppose instead that the response variable is **categorical** (actually whether the majority of chromatophores in a study of the

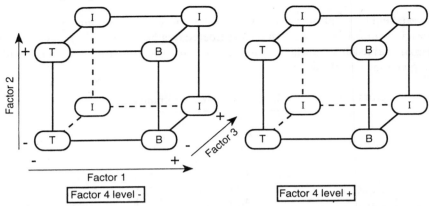

Figure E.2 Response diagram for Exercise E.6(a). Reproduced with permission from Sundberg (1994).

regeneration of a coenzyme are in the bottom phase (B), the top phase (T) or in an intermediary layer (I)). The results are shown in Fig. E.2, where + and − denote the high and low levels, respectively, of each factor.

How do the four factors influence the response (i.e. the location of the chromatophores?) You will not know how to do a 'standard analysis' (if one exists), so use your common sense. Refer to the factors as factor 1, factor 2, etc. (Of course in the chemometric context, you would need to interpret the results by referring to the names and levels of the factors.)

(b) In a pharmaceutical company, the standard method for determining the amount of an active agent in a type of skin cream had given unreliable results for some time and was therefore investigated with a 2^4 experiment. The four factors examined were: (A) pH value (low and high); (B) soaking time (low and high); (C)

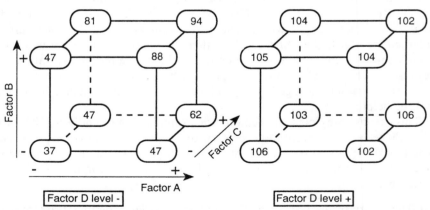

Figure E.3 Response diagram for Exercise E.6(b). Reproduced with permission from Sundberg (1994).

temperature (low and high); and (D) age of enzyme (old and new). The response variable was the yield expressed as the percentage of the predicted value (with a higher value being better). The results are shown in Fig. E.3, where + denotes the high levels of factors A, B and C and denotes new enzyme for factor D.

How do the four factors influence the response. Which one is most important? Do the factors interact? Summarize the results.

Notes on Exercise E.1

The easiest question to answer is (c) – there is no rule which says you have to answer the questions in the order presented! All the row and column sums are integers. This is so unlikely that it is clear that the data are fictitious. They are in fact adapted from an artificial exercise in a textbook on design. No background information was given and the units of measurement were not stated. Any analysis will be what Professor D.J. Finney has called an analysis of numbers rather than data – a useful distinction. Nevertheless, the numbers given in Table E.1 provide a useful tutorial exercise in assessing row and column effects.

The answer to (a) is clearly meant to be a randomized block design, although no mention of randomization is made in the question. In practice you would want to find out if randomization really had been used (as well as getting background information on units of measurement, objectives, etc.!).

When analysing the results of a randomized block experiment, the first step is to calculate the row and column means (which have already been added to Table E.1), and then carefully examine them. In this case the differences between column means are relatively large, and those between row means relatively small. While it may be hard to assess the residual variation without removing the row and column effects, I hope that it is intuitively obvious to you in this case that the treatment means are significantly different, but that block means are not. Of course the significance of block effects is not very important as blocks are used to control variation and there is usually relatively little interest in the size of block differences, except insofar as they affect the efficiency of the design (was it worth blocking in the first place?).

The 'proper' analysis here would be to carry out a two-way ANOVA which partitions the total variation into components due to variation between treatments (columns), between blocks (rows) and the residual variation, giving the results shown in Table E.2. The F-ratio for treatments is 49.6 ($= 480/9.67$) on 7 and 14 DF and is highly significant. In contrast, the F-ratio for blocks is not significant.

As the results of these significance tests are really obvious in advance, the main use of the ANOVA here is to provide an unbiased estimate of the residual variance (or mean square), namely $s^2 = 9.67$. This may be useful in a variety

Table E.2 The ANOVA table for the data of Table E.1

Source	Sum of squares	Degrees of freedom	Mean square	F
Treatments	3360.0	7	480	49.6
Blocks	48.0	2	24	2.5
Residual	135.36	14	9.67	
Total	3543.36	23		

of follow-up procedures. In particular, having found a significant difference between treatments, the ANOVA would normally be followed by an examination of the treatment (column) means to see which is 'best'. Unfortunately, we do not know if a high or low observation is 'good' in this case, and so we simply note that treatments 2 and 5 give the highest results and treatment 4 the lowest. The standard error of the difference between two treatment means is $\sqrt{[9.67(\frac{1}{3} + \frac{1}{3})]} = 2.5$, and so the extreme treatment means clearly differ significantly from the next highest or lowest treatment mean as appropriate.

Illustrative commands for carrying out the ANOVA using the MINITAB and GLIM3 packages are given in Appendix B. The output from MINITAB is much easier to understand.

Moral

With a one-way or two-way ANOVA, the results of any significance tests may be obvious in advance, and the main purpose of the ANOVA may then be to provide an estimate of the residual variance which can be used to estimate confidence intervals for the differences between treatment means.

Solution to Exercise E.2

These data are taken from Chatfield (1983, p. 256) where they were presented as a technique-oriented exercise with the instruction 'Test the hypothesis that there is no difference between burners'. Although the accompanying text in the above book makes clear that a full statement of the results of an analysis is required, rather than just saying if a result is significant or not, the brief solution provided in the answers section simply says that '$F = 19.8$ is significant at the 5% level'. This tells us little or nothing in isolation.

In the spirit of this book, the instructions for this example are quite different, namely to 'analyse the data'. You should therefore concentrate on understanding the data and aim for a more mature approach in which the ANOVA is relegated to a more subsidiary position.

These data are more highly structured than those in Example E.1 in that each observation is indexed by a particular row (day), a particular column

Table E.4 Mean values for burner efficiency data

Day 1 = 17.7	Engine 1 = 15.7	Burner 1 = 14.3
Day 2 = 17.3	Engine 2 = 16.7	Burner 2 = 15.3
Day 3 = 13.3	Engine 3 = 16	Burner 3 = 18.7
	Overall mean = 16.1	

(engine) and a particular burner. It is the elegance and power of the Latin square design which enables us to separate the effects of these three factors and examine them all at the same time. Presumably the comparison of burners is of prime importance. This is difficult to do 'by eye' because of the Latin square design. It is therefore sensible to begin by calculating row, column and burner (treatment) means, as shown in Table E.4.

Note that the day 3 mean is rather low and the burner 3 mean is rather high. It is still difficult to assess if these differences are significant. To assess the size of the residual variation we can remove the row, column and treatment effects to calculate the table of residuals. With an orthogonal design, this is easy to do by subtracting appropriate means. For example, the residual for the top left observation in Table E.3 is given by

$$(16 - 16.1) \quad - (17.7 - 16.1) - (15.7 - 16.1) - (14.3 - 16.1) = 0.5$$

$$\text{(Residual from} - \text{(Day effect)} - \text{(Engine effect)} - \text{(Burner affect)}$$
$$\text{overall mean)}$$

The full list of residuals (reading by rows and calculated more accurately to two decimal places) is: 0.56, -0.44, -0.11, -0.11, 0.56, -0.44, -0.44, -0.11, 0.56. (Check that they sum to zero.) These residuals look small compared with the differences between burner means and between day means. However, there are only two degrees of freedom left for the residuals so that they are likely to appear to have smaller variance than the true 'errors'. It seems that there is no substitute for a proper ANOVA in this case to determine the significance of the burner effect. Using GLIM (Example B.3 in Appendix B), Table E.5 shows the resulting ANOVA table. The F-ratio for burners looks high, but because the degrees of freedom are small, the result is only just significant at the 5% level. We conclude that there is some evidence of a difference between

Table E.5 ANOVA table for the data in Table E.3

Source	SS	DF	MS	F
Burners	30.88	2	15.44	19.8
Days	34.89	2	17.44	22.3
Engines	1.56	2	0.78	1.0
Residual	1.56	2	0.78	
Total	68.89	8		

burners, and in particular burner 3 looks best. It is helpful to use the residual MS to calculate the standard error of the difference between two burner means, namely $\sqrt{[0.78(\frac{1}{3} + \frac{1}{3})]} \simeq 0.7$. We see that burner 3 is nearly five standard errors above the next burner.

Of course this is a very small experiment and we should not read too much into it. Indeed, you may think this a rather trivial data set. However, its small scale enables us to appreciate the problems involved, and I have deliberately chosen it for that reason. The reader could now go on to more complicated data sets such as the replicated (4 × 4) Latin square in data set 9 of Cox and Snell (1981). There it is necessary to take out a replicate (day) effect on one DF as well as row, column and treatment effects.

Moral

With a complicated design like a Latin square, an ANOVA may be indispensible in assessing the significance of the treatment effects and in providing an estimate of the residual variance for use in estimation.

Notes on Exercise E.3

(a) Regarding the wheat varieties as 'treatments' and the centres as 'blocks', and assuming that plots are allocated randomly to treatments at each centre, then we have a form of randomized block experiment. However, the experiment is incomplete as not all varieties are tested at each centre. Moreover, the experiment is unbalanced as the numbers of observations on each variety may differ. Unbalanced, non-orthogonal designs were generally avoided for many years as they lead to a more complicated analysis. However, the computing resources now available make this much less of a problem.

Another drawback to unbalanced designs is that comparisons between different varieties may have different precisions. However, this will not be too much of a problem unless the design is 'highly unbalanced'. It is hard to say exactly what this means but if there are, say, four times as many observations on one treatment as another, or if different pairs of treatments occur several times or not at all in the same block, then there may be difficulties. Reasonably sensible unbalanced designs are now increasingly used when practical problems make balanced designs difficult or impossible to achieve.

(b) Let us start as usual by looking at the data. We see that there are large differences between centres (e.g. compare E8 with N10). This is to be expected given that crops generally grow better in a warmer climate. There are also some systematic, though smaller, differences between varieties. In particular, we note that the Mardler variety gives the highest yield at seven of the 10 centres.

It may help to calculate the row and column means, as given in Table E.7, and add them to Table E.6, although, with an unbalanced design, it should be

Table E.7 (Unadjusted) row (a) and column (b) means of wheat yields from Table E.6

(a)

Variety	Mean
Huntsman	5.73
Atou	6.03
Armada	6.40
Mardler	6.78
Sentry	6.98
Stuart	7.16

(b)

Centre	Mean
E1	6.07
E2	6.81
N3	4.99
N4	4.79
N5	5.62
N6	6.27
W7	7.25
E8	8.12
E9	7.47
N10	4.38

borne in mind that they are potentially misleading. For example, the Stuart variety gets the highest row mean, but three of its four readings are at three of the 'best' centres and so its row mean is artificially high. It may also help to plot the data as in Fig. E.1, where the centres have been reordered by region.

Given the external estimate of precision, it is quite clear that there are major differences between centres, and smaller, though non-negligible, differences between varieties. Bearing in mind that the Sentry and Stuart varieties were only tested at centres 7–10, the choice of 'best' variety seems to lie between Mardler, Stuart and Sentry. Finally, we ask if there is any evidence of interaction, by which we mean whether some varieties grow particularly well (or badly) at certain centres. There is some mild evidence of interaction (e.g. Huntsman does rather well at N3), but the effects are generally small and the lines in Fig. E.1 seem reasonably parallel. There are certainly no obvious outliers.

The above analysis may well be judged adequate for many purposes. However, it would be helpful to have a 'proper' comparison of varieties which takes the unbalanced nature of the design into account. We therefore present the results of an analysis using GLIM. As the effects of variety and centre are not orthogonal, we can get two ANOVA tables depending on whether the variety or centre effect is fitted first (Table E.8). Although 'different' in the top two rows, the two ANOVAs proclaim the same general message. The centre effect gives a much larger F-ratio than the variety effect, but the latter is still significant at the 1% level whether it is fitted first or second.

The GLIM analysis assumes normal errors with homogeneous variance. The latter assumption is not unreasonable given the ranges of values at the different centres. Other assumptions can be checked in the usual way by looking at the residuals. One problem with an unreplicated two-way design is that it is difficult

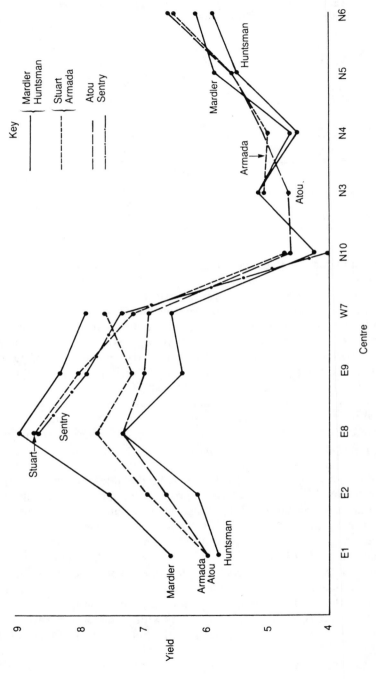

Figure E.1 Yields of winter wheat at 10 centres.

Table E.8 ANOVA tables for wheat yield data: (a) fitting varieties first, (b) fitting centres first

(a)

Source	SS	DF	MS	F
Varieties	10.67	5	2.13	13.6
Centres after varieties	60.85	9	6.76	43.1
Residual	4.87	31	0.157	
Total	76.39	45		

(b)

Source	SS	DF	MS	F
Centres	67.14	9	7.46	47.5
Varieties after centres	4.38	5	0.88	5.6
Residual	4.87	31	0.157	
Total	76.39	45		

or impossible to tell if a large residual indicates an outlier or is caused by an interaction term. With only one replicate, it is not possible to fit all interaction terms and also get an estimate of the residual variance. For the model fitted in Table E.8, the residual standard deviation is $\sqrt{0.157} \simeq 0.4$, which is larger than the external estimate of precision, namely 0.2. However, if it were possible to take interactions into account, the value of 0.4 would probably be reduced.

The GLIM analysis also provides estimates of the row and column means adjusted to take account of the unbalanced design. For the varieties, if we leave the adjusted means for Huntsman and Atou unchanged, then Armada and Mardler decrease by 0.2 (because they are not included at the 'worst' N10 centre), while Sentry and Stuart decrease by 0.57. Thus the adjusted means for Mardler and Stuart are nearly identical and better than the rest. Given that Mardler is best at seven centres and has been tried more extensively, I would prefer Mardler, other things being equal.

For the centres, if we leave the adjusted means for the first six centres unchanged, then centres 7–9 decrease by about 0.12, while centre 10 decreases by 0.06. Thus centre 8 gives the best results and centre 10 the worst.

An alternative, more complicated, analysis is given by Patterson and Silvey (1980) and should certainly be considered by the specialist agricultural statistician. Other readers may find that analysis rather opaque and may well find the analysis given here not only easier to follow, but also easier to interpret and more suitable for providing a useful summary of the data.

Moral

Even with an unbalanced design, an exploratory examination of the data may still be successful in highlighting their main features.

Discussion of Exercise E.4

Two observations have been made at every possible combination of the different selected levels of the four variables, and this is called a twice-replicated $(6 \times 2 \times 2 \times 4)$ **complete factorial experiment**. The total number of observations is equal to $2 \times 6 \times 2 \times 2 \times 4 = 192$. An important question not answered in the statement of the problem is whether the design has been randomized, both in the allocation of the rigs and in the order of the experiments. In fact the design was not properly randomized, although it was at least not fully systematic. However, we will proceed as if the data were reliable, while bearing this drawback in mind.

Before starting the analysis, it would be important in practice to get more background information and clarify the objectives precisely. Do we want a model describing behaviour over the whole range of experimental conditions or are we mainly interested in maximizing failure time? We concentrate on the latter.

You may not have seen data like these before. Nearly a quarter of the observations are listed as NF, which means that there was no failure in the first 20 hours, at which time the test was stopped. Data like these are called **censored** data (see also Example B.5). What do we do about them? It might be possible to analyse the quantitative information in the lower part of Table E.9, which is almost unpolluted by censored values, but this is the least interesting part of the data given our interest in maximizing failure time.

The physicist who brought me the data had arbitrarily inserted the value 22 hours for each NF value and then carried out a four-way ANOVA. This suggested that the main effects of load, promotor and charging potential are significant while pre-charging is not. However, this approach is obviously unsound, not only because censored values should not be handled in this way but also because inspection of the data suggests that the 'error' distribution is not normal. Rather it is skewed to the right with many observations between 0 and 5, relatively few between 5 and 20 and rather more exceeding 20. Looking at paired replicates, it is also clear that the residual variance is not constant. A distribution such as the gamma or Weibull distribution is indicated.

Although the data are highly structured experimental data, you may have little idea how to proceed. Rather than give up, you should approach the data as in Chapter B by exploring them using common sense. Indeed, perhaps this example should be in Chapter B.

Why not start in a simple-minded way by treating the data as binary (fail or not-fail), even though this effectively throws away some information. The main effects of the four factors can be assessed by finding the frequencies of not-fails at each level of each of the factors. For example, the frequencies at the four levels of charging potential are given in Table E.10, and it can be seen that the observed frequencies are obviously significantly different, a view which can (and should? – see below) be confirmed by a χ^2 goodness-of-fit test. Similarly, we

Table E.10 Frequencies of NFs at different levels of charging potential

	Charging potential (V)			
	−0.7	−0.9	−1.2	−1.45
Frequency of not-fail	28	7	5	5

find that the main effects of load and promotor are significant while pre-charging is not. In fact these results are the same as produced by the dubious four-way ANOVA mentioned above and it appears that the three significant main effects are so obvious that virtually any method will reveal them.

The appropriate χ^2-test for data such as those in Table E.10 may not be obvious. It is necessary to construct the full two-way table, including the frequencies of failures, as in Table E.11, since the significance of the differences between not-fails depends on the relative sample sizes. The null hypothesis (H_0) is that the probability of a not-fail (or equivalently of a fail) is the same at each level of charging potential. There is a special test for testing equality of proportions but the standard χ^2-test is equivalent and easier to remember. As column totals are equal, the expected frequencies in the four cells of the top row of Table E.11 are all 11.25, while the expected frequencies in the bottom row are all $(48 - 11.25) = 36.75$. Then $\chi^2 = \Sigma[(\text{obs.} - \text{exp.})^2/\text{exp.}] = (28 - 11.25)^2/11.25 + \cdots = 43.7$. The distribution of the test statistic under H_0 is χ_3^2 where DF = (No. or rows − 1) × (No. of columns − 1) as usual for a two-way table. Since $\chi_{3,0.01}^2 = 11.34$, we have strong evidence to reject H_0. Indeed, the contribution to the χ^2-statistic from the top left-hand cell is so large by itself that there is really no need to calculate the rest of the statistic. By inspection, we see that failure time can be increased by decreasing load and promoter and by increasing charging potential. Put more crudely, the values of the factors which maximize failure time are in the top left-hand corner of Table E.9, and this seems obvious on re-examining the table. It seems unlikely that the lack of randomization will affect these conclusions.

The above analysis may well be judged adequate for many purposes, although it makes no attempt to assess interaction effects. However, suppose you wanted to construct a formal model using full information from the censored data. This

Table E.11 Frequencies of NFs at different levels of charging potential

	Charging potential (V)			
	−0.7	−0.9	−1.2	−1.45
Frequency of not-fail	28	7	5	5
Frequency of fail	20	41	43	43

is a much more sophisticated task and it is unlikely that you will know how to proceed. Neither did I when first shown the data! By asking various colleagues, I eventually discovered that a standard solution is available for these sort of data using a log-linear proportional hazards model and a censoring variable (e.g. Appendix A.15 and Aitkin and Clayton, 1980). The data can then be analysed in a relatively straightforward way using the GLIM computer package. The details are outside the scope of this book.

Moral

When faced with data of a type not seen before, 'play around' with them in order to get an idea of their main properties. Do not be afraid to ask other statisticians for help if this proves necessary.

Discussion of Exercise E.5

The instructions to 'analyse the data' are typically vague and open-ended, as they often are in real life. What is clear is that the main comparison of interest is between treatments A and B and that we want to end up with a clear summary of what the data are saying. Begin, as always, with an IDA. Calculate summary statistics for the four groups of observations (e.g. mean and standard deviation, or median and interquartile range) and also plot a boxplot or dotplot of each group with the same scale so that the four plots can readily be compared. It will then be clear that (i) there is a large difference in the effect of treatments A and B (average mean difference is 1.27); (ii) there is little or no difference between labs I and II (average mean difference is 0.03); (iii) the average difference between A and B is larger in lab II than in lab I (1.78 rather than 0.76) indicating some possible treatment–lab interaction; (iv) there is no sign of any differences in the scatter of the four groups (the within-group standard deviations only vary from 0.67 to 0.97). The lung weights under treatment A are substantially heavier on average than those for treatment B, as might be expected on biological grounds. Animals kept in an atmosphere containing lots of dust can be expected to suffer from a deteriorating chest condition, of which increased weight is a symptom, and this may in turn lead to tuberculosis.

Is any further analysis indicated? While it looks fairly clear that the treatment effect is 'significant', it is not obvious from the IDA whether the interaction effect is significant, and it could also be helpful to get an estimate of the residual variation in order to get a confidence interval for the difference in the effect of treatments A and B. For this nicely balanced data set, it is easy to carry out an ANOVA which shows, as expected, that the difference between A and B is highly significant ($F = 16.9$). However, the lab effect is nowhere near significant as expected (F-ratio much less than one) while the interaction effect is also not significant ($F = 2.64$ on 1 and 28 DF). The estimate of the residual variance is

0.771 on 28 DF. A thorough model-based analysis will also include a residual analysis, checks on assumptions, etc., but this will not be described here. Rather a summary of the data is, as noted above, that lung weights of rats kept in a dusty atmosphere are significantly heavier on average than those of rats kept 'dust-free'.

Moral

With a highly replicated experimental design, it is worth plotting a boxplot or a dotplot of the group of observations in each cell of the design. The results may then be obvious without the need for an ANOVA except perhaps to get an estimate of residual variance, though in this case the ANOVA was also useful to assess the interaction.

Discussion of Exercise E.6

(a) There are no numbers to analyse, so let us examine Fig. E.2 using common sense. It is apparent that factors 2 and 4 do not show any influence on the observed responses for the given experimental conditions, since the categorical response is identical for every combination of factors 1 and 3 when factors 2 and/or 4 are changed from the low to the high level. When factor 3 is at high level, factor 1 has no influence since the response is I whatever the value of factor 1, whereas when factor 3 is at its low level, the response changes from T to B when factor 1 changes from low to high. This means that the effects of factors 1 and 3 are not **additive**, but rather **interact**.

One way to summarize the results is as follows: factors 2 and 4 have no effect on the response variable under the range of conditions examined, whereas factors 1 and 3 do have an effect in an interactive way. The response is always I at the high level of factor 3 (where factor 1 also has no effect), but may be T or B at the low level of factor 3 depending on whether factor 1 is low or high.

(b) Here we do have some numbers, but the results are so clear-cut that no formal analysis is really required. It is clear from Fig. E.3 that there is a drastic difference between the effects of the two enzyme types (old and new). With the new enzyme, the response is essentially constant at a value slightly higher than 100 (which is 'good'), whereas with the old enzyme, the response is generally much lower and varies with the other three factors. Sundberg (1994) rightly says that 'No further analysis is really required; it appears sufficient to make sure that in future the enzyme be fresh'.

Sundberg does in fact go on to take a closer look at the estimates of the different effects for pedagogical interest. Taking the usual definitions of main effects and interactions (note that the GLIM package uses different definitions), D gives by far the largest main effect (of course), while the only substantial first-order interactions are those between A and D, B and D, and C and D. All the latter turn out to be of similar size to the corresponding main effect (other than D), but with opposite sign (e.g. the estimate of A main effect is approximately minus the estimate of the AD interaction). This simply reflects the fact that the responses are nearly constant at the high level of D (think about it!).

Looking at Fig. E.3 again, we see that the response is best for the old enzyme when factors A, B and C are all at high level. This suggests that, if there are no disadvantages associated with the choice (e.g. cost), then high values of A, B and C should generally be used as this would allow for the possibility that the freshest enzyme available has started to age, and would also cope with the situation when the old enzyme is the only type available or is used by mistake. However, the main recommendation is to use the freshest source of enzyme possible.

Moral

This exercise demonstrates once again the need to **look at data** and be prepared to modify standard procedures to cope with non-standard situations, namely handling a categorical response variable and realizing the implications of the high, near-constant values of the response variable at the high level of D.

F
Time-series analysis

A time series is a collection of observations made sequentially through time. One special feature of time-series data is that successive observations are usually known **not** to be independent, so that the analysis must take into account the **order** of the observations. As time-series analysis constitutes my particular research speciality, I cannot resist including a short section of time-series problems (see also Exercise G.2). These exercises should not be attempted unless you have studied a basic course in time-series analysis (except perhaps Exercise F.1).

The first step, as usual, is to clarify the objectives of the analysis. Describing the variation, constructing a model and forecasting future values are three common aims. The second important step is to plot the observations against time to form what is called a **time plot**. The construction of a time plot is the main ingredient of IDA in time-series analysis and should show up obvious features such as trend, seasonality, outliers and discontinuities. This plot is vital for both description and model formulation and, when effects are 'obvious', may well render further analysis unnecessary.

Apart from the time plot, the two main tools of time-series analysis are the **autocorrelation function** and the **spectrum** (Appendix A.14). Time-series modelling based primarily on the autocorrelation function is often called a **time-domain** approach. Alternatively, with engineering and physical science data in particular, a **frequency-domain** approach based on spectral analysis may be appropriate, but this will not be illustrated here.

Exercise F.1 Forecasting sales data

Figure F.1 shows the number of new insurance policies issued by a particular life office in successive months over seven years. You have been asked to produce forecasts for the next 12 months. How would you set about answering this question?

Exercise F.2 Forecasting TV licence numbers

The numbers of TV licences known to be currently valid in the UK, in successive

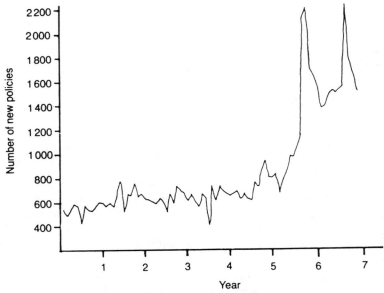

Figure F.1 Numbers of new insurance policies issued by a particular life office.

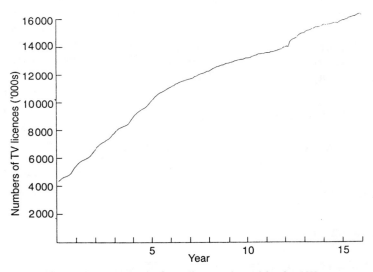

Figure F.2 Numbers of TV licences issued in the UK.

Table F.1 Numbers of TV licences (in '000s) issued in the UK in successive months

Year 1	4 308	4 407	4 504	4 581	4 624	4 676
	4 726	4 786	4 884	5 078	5 262	5 400
Year 2	5 539	5 649	5 740	5 812	5 863	5 922
	5 980	6 044	6 140	6 291	6 433	6 570
Year 3	6 757	6 863	6 966	7 050	7 119	7 170
	7 270	7 331	7 398	7 524	7 657	7 761
Year 4	7 898	7 995	8 090	8 147	8 201	8 253
	8 295	8 345	8 424	8 571	8 731	8 899
Year 5	9 044	9 153	9 255	9 347	9 413	9 495
	9 550	9 628	9 718	9 844	9 987	10 114
Year 6	10 220	10 368	10 470	10 569	10 647	10 702
	10 753	10 817	10 880	10 963	11 028	11 076
Year 7	11 149	11 187	11 286	11 321	11 392	11 441
	11 485	11 522	11 553	11 602	11 634	11 657
Year 8	11 693	11 745	11 834	11 866	11 930	11 984
	12 040	12 075	12 110	12 167	12 224	12 231
Year 9	12 290	12 375	12 443	12 484	12 542	12 570
	12 622	12 661	12 664	12 731	12 778	12 789
Year 10	12 830	12 863	12 885	12 944	12 967	13 010
	13 026	13 057	13 061	13 097	13 146	13 155
Year 11	13 161	13 182	13 253	13 296	13 336	13 358
	13 436	13 428	13 448	13 455	13 489	13 516
Year 12	13 502	13 506	13 567	13 586	13 606	13 641
	13 674	13 716	13 755	13 782	13 888	13 919
Year 13	13 960	13 908	14 267	14 392	14 463	14 510
	14 556	14 687	14 776	14 862	14 880	14 910
Year 14	15 016	15 068	15 093	15 136	15 202	15 230
	15 315	15 331	15 324	15 377	15 399	15 506
Year 15	15 439	15 488	15 509	15 528	15 559	15 595
	15 623	15 630	15 576	15 698	15 770	15 809
Year 16	15 831	15 381	15 899	16 000	16 024	16 075
	16 100	16 124	16 183	16 254	16 292	16 188

months over 16 years, are given in Table F.1 and plotted in Fig. F.2. How would you set about producing forecasts for the next 12 months?

Exercise F.3 Time-series modelling

Figure F.3 shows three time series which exhibit widely different properties. Discuss the sort of time-series model which might be applicable to each series. Comment also on the quality of the three time plots.

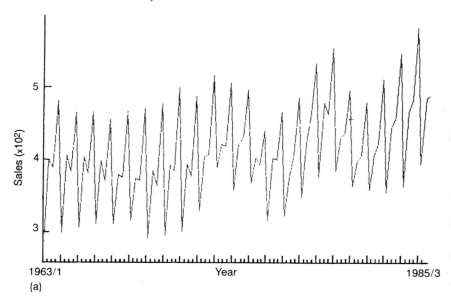

(a)

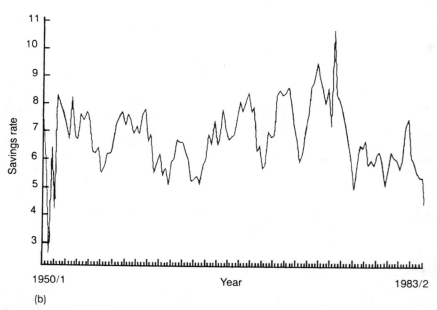

(b)

Figure F.3 Quarterly time-series data: (a) sales of a particular type of shoe; (b) California savings rate.

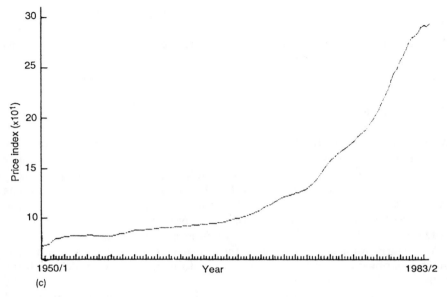

Figure F.3 Quarterly time series data: (c) California consumer price index.

Discussion of Exercise F.1

In view of the sudden changes in sales towards the end of the series, you should not be tempted simply to apply a standard projection technique. Indeed, you may think that you have been asked a silly or impossible question, and you would be correct. Nevertheless, the author was asked this question, and so the aspiring statistician needs to learn how to deal with silly or impossible questions.

I began by asking what had produced the sudden increases in sales figures. It turned out that there had been a sales drive towards the end of year 5 and another towards the end of year 6. The next obvious question is whether further sales drives are planned for the future and whether sales are expected to continue at a higher level anyway. Any forecasts are crucially dependent on the answer. While the statistician may still be able to offer some assistance, it is clear that informed guesswork is likely to be preferable to statistical projections and one should not pretend otherwise.

This example illustrates the fact that the time plot is an indispensable stage of a time-series analysis. The abrupt changes visible in Fig. F.1 tell us that we cannot expect any univariate model adequately to describe the whole series. The time plot is therefore invaluable both for describing the data, and also for model formulation, albeit in a negative way in this example.

Moral

Always start a time-series analysis by plotting the data, and asking questions to get background information.

Discussion of Exercise F.2

As in Exercise F.1, the time plot of Fig. F.2 is an invaluable part of the analysis, though in a completely different way. The series in Fig. F.2 is much more regular than that in Fig. F.1 and we expect to be able to produce reliable forecasts.

There are three features to note in Fig. F.2. First, note the upward trend as more TVs are bought. Second, note the small but noticeable seasonal variation in the first five years' data. This seasonality appears to vanish thereafter. Third, note the small but noticeable jump in mean level around year 12.

We clearly want a forecasting method which can cope with a trend. It is much less obvious as to whether it is better to fit a seasonal model to all the data or to fit a non-seasonal model to the data from year 5 onwards. The method we choose must also be able to deal with the change in mean around year 12.

There is no 'best' forecasting method. Rather you should choose a 'good' method which you understand and for which computer programs are available. My own favourite for this type of data is the Holt–Winters forecasting procedure (e.g. Chatfield, 1989) which is based on fitting a model containing a local level, a local trend, seasonal terms (either additive or multiplicative) and an error term. The level, trend and seasonal terms are all updated by exponential smoothing. When applied to these data, I found that much better forecasts of the last two years are obtained by fitting a non-seasonal model to the latter part of the data, rather than fitting a seasonal model to all the data. The possibility of ignoring the first part of the data (which horrifies some people!) arises directly from the time plot.

Moral

There are sometimes good reasons for ignoring part of a data set, particularly if external conditions have changed or it is otherwise unreasonable to expect a single model to describe all the data.

Notes on Exercise F.3

No single class of models can deal satisfactorily with all the different types of time series which may arise. Some subjective judgement must be used.

Series (a) is typical of many sales series in that there is a high seasonal component. In nearly every year sales are lowest in the first quarter of the year and highest in the final quarter. There is also some evidence of trend as sales increase towards the end of the series. A good time-series model should explicitly model this trend and seasonal variation which forms a very high proportion of the total variation. Thus a trend-and-seasonal model is preferred to a Box–Jenkins ARIMA model, where model fitting would involve differencing away the trend and seasonality and devoting most effort to modelling the autocorrelation in the resulting differenced series. The latter is relatively unimportant for this series.

In contrast, series (b) has no seasonal component and little trend and yet there are short-term correlation effects which do need to be modelled. An ARIMA model might well be fruitful here, particularly if the first year's atypical observations are removed.

Series (c) is quite different again. Exponential growth characterizes many economic indices. This exponential growth needs to be explicitly modelled, perhaps after taking logarithms. Neither a trend-and-seasonal model, nor an ARIMA model, is appropriate for describing the raw data.

The time plots in Fig. F.3 are given exactly as they came out of the computer (which is generally not advisable). Note the poor labelling of the horizontal axes which makes it virtually impossible to tell which year an observation came from.

Moral

Different classes of time-series model are appropriate for different types of series.

G
Miscellaneous

This chapter comprises seven exercises which are rather varied in nature and cannot be readily classified elsewhere. They range from a probability problem (G.1) through a variety of statistics problems to the improvement of communication skills.

Exercise G.1 Probability and the law

In a celebrated criminal case in California (People versus Collins, 1968), a black male and a white female were found guilty of robbery, partly on the basis of a probability argument. Eyewitnesses testified that the robbery had been committed by a couple consisting of a black man with a beard and a moustache, and a white woman with blond hair in a ponytail. They were seen driving a car which was partly yellow. A couple, who matched these descriptions, were later arrested. In court they denied the offence and could not otherwise be positively identified.

A mathematics lecturer gave evidence that the six main characteristics had probabilities as follows:

negro man with beard	1/10
man with moustache	1/4
girl with ponytail	1/10
girl with blond hair	1/3
partly yellow car	1/10
inter-racial couple in car	1/1000

The witness then testified that the product rule of probability theory could be used to multiply these probabilities together to give a probability of 1/12 000 000 that a couple chosen at random would have all these characteristics. The prosecutor asked the jury to infer that there was only one chance in 12 million of the defendants' innocence, and the couple were subsequently convicted.

Comment on the above probability argument.

Table G.1 Numbers of daily admissions to a psychiatric hospital in a one-year period, starting on Sunday 1 January. The lines indicate the end of a week and totals are also given

January	wk	March	wk	May	wk	July	wk	September	wk	November	wk
0		1		3		1		1	18	1	
1		3		2		3		0		3	
4		1	14	3		2		5		2	14
3		0		5		1		1		0	
3		1		2	19	0		2		4	
7		9		1		1		7		0	
3	21	2		1		4	12	2		3	
0		3		2		2		0	17	3	
2		4		2		3		0		7	
1		0	19	3		0		0		1	18
4		1		5		3		1		1	
3		2		0	14	7		3		3	
6		3		0		3		6		5	
1	17	1		1		2	20	4		2	
0		1		1		1		3	17	4	
1		3		3		3		1		8	
5		0	11	2		5		3		1	24
3		0		1		3		4		1	
6		1		2	10	5		3		3	
4		3		1		5		1		3	
0	19	1		2		2	24	7		1	
0		2		4		3		3	22	2	
3		1		1		2		0		2	
0		2	10	3		3		2		4	16
2		2		7		2		1		0	
3		1		2	20	5		3		3	
5		2		1		2		2		4	
4	17	4		1		0	17	4		2	
0		1		1		0		0	12	3	
4		2		0		0		1		7	
2		0	12	3		3					

February	wk	April	wk	June	wk	August	wk	October	wk	December	wk
3		1		3		4		1		1	20
2		8		2	11	4		5		2	
4		2		0		1		1		4	
1	16	3		4		2	14	3		1	
2		4		2		2		2		2	
0		4		3		2		2	15	7	
2		1	23	3		3		1		3	
3		1		4		3		2		2	21
6		3		2	18	2		2		1	
2		4		2		4		1		0	
1	16	1		4		5	21	4		5	
2		3		0		1		1	13	4	
1		2		5		2		1		4	
1		1	15	6		3		1		3	
1		0		4		6		3		2	19
3		5		2	23	3		2		0	
2		8		1		5		3		1	
2	12	1		4		4	24	2	13	3	
1		3		3		1		1		3	
3		2		4		0		2		4	
0		3	22	2		3		5		1	
6		2		2		3		0		2	14
3		0		1	17	3		4		1	
5		4		2		2	14	2		0	
2	20	8		3		1		4	18	0	
1		1		0		1		0		2	
2		2		3		3		3		3	
3		2	19	5		4		1		2	8
3		2		4		3		4		3	
		2		1	18	5				4	

Exercise G.2 Hospital admissions – communication skills

This exercise requires you to reply to the following letter seeking advice:

Dear Statistician,

The data in Table G.1 show the numbers of new patients arriving at a psychiatric hospital during a recent one-year period. We are interested in finding out if there are any systematic variations in arrival rate, especially any that might be relevant in planning future operations, as we are currently reconsidering the running of this unit. We are particularly interested in any evidence of cyclic behaviour. Please analyse the data for us.

Yours sincerely,

The Assistant Planning Officer
Area X Regional Health Authority

The letter which you write in reply should not exceed three sides of paper, and should be written for a numerate layman rather than a statistician. It should be accompanied by tables and graphs as appropriate, covering not more than four sides.

Exercise G.3 Testing random numbers

In a national lottery, 125 ticket numbers were drawn by computer in one prize draw. The final six digits of each number were supposed to be random. These values are tabulated in Table G.3. Are the numbers random?

Table G.3 The final six digits of the 125 winning prize numbers in a national lottery

535850	842420	655257	469227	885878
603715	863855	754258	883261	571046
075779	048633	111337	346576	051352
724004	089507	552867	476843	025348
355865	001250	095391	934011	094093
771594	616635	992135	473416	021096
726862	768318	218966	928474	538201
593721	318619	908649	198296	122079
081625	046970	477814	516512	738317
461645	489085	619015	627585	222443

Table G.3 continued

513613	937397	761133	117830	726151
677301	573766	432414	545267	428765
982835	079759	394619	082633	711609
149985	944572	002723	077330	328214
466322	573243	212831	886922	233534
676237	303859	308467	796715	264418
178050	058331	287017	929288	224397
410495	742978	499739	544181	445026
797412	628442	634667	855302	327110
729083	195838	151508	801377	760187
347732	087664	513146	809011	650496
317829	922459	535252	196620	038503
569158	266110	952900	484856	413744
940045	029951	876305	240466	621546
427035	775419	099017	631855	906895

Exercise G.4 Sentencing policy – two-way tables?

A study was carried out on consecutive groups of men and women who were convicted of theft and shoplifting in an English city. A total of 100 females and 100 males were examined and their sentences were classified as lenient or severe according to the average sentence given for the person's particular crime. The results were as follows:

	Lenient sentence	Severe sentence	Total
Male	40	60	100
Female	60	40	100

Are females treated more leniently than males? If they are, speculate as to why this should be so.

Exercise G.5 Dose response data

(a) Patients with a particular medical condition are treated by a new drug which unfortunately has some undesirable after-effects. It is thought that the chance of getting after-effects may depend on the dose level, x, of the drug. At each dose level, the number of patients suffering after-effects is noted from a batch of recent case histories giving the following data:

Dose level, x_i	0.9	1.1	1.8	2.3	3.0	3.3	4.0
No. of patients, n_i	46	72	118	96	84	53	38
No. with after-effects, r_i	17	22	52	58	56	43	30

Analyse the data in whatever way you think is appropriate. Estimate the ED_{50}, which is the dose level giving after-effects in 50% of patients.

(b) An investigation was carried out into the effect of a new drug being tested for its toxicological effects. Twenty rats were each given a specified dose of the drug at five dose levels. As this was a screening experiment, a wide dose range was investigated. The results were as follows:

Dose level	1	2	3	4	5
Sample size	20	20	20	20	20
No. of dead rats	0	0	1	20	20

Analyse the data and estimate in particular the dose level at which half the rats will be killed (called the median lethal dose or LD_{50}).

Exercise G.6 Using a library

(a) *Finding a particular statistic*   A wide range of published official statistics are available in many libraries. A statistician should know how to find them if necessary. Find the most recent published value of one (or more) of the following quantities for the country in which you live:

   (i) The total number of births.
   (ii) The total number of deaths in road accidents.
   (iii) The total number of university students (full-time undergraduates only, or what?).
   (iv) The number of unemployed.
   (v) The average number of hours per week spent watching TV.

You should specify where the statistic was found, say exactly what the statistic measures, and specify the appropriate year or part of year.

(b) *Using statistical journals*   A statistician should not be afraid of looking up recent research in statistical journals or using index journals. Listed below (in alphabetical order) are the bibliographic details of some general-interest papers published in the last 15 years or so. Select a title which interests you, find the relevant journal in your library, make a photocopy of the paper, read it carefully, and write a critical summary of it in not more than about five sides.

Using the *Science Citation Index*, or any other aid, find some (not more than three) of the journal articles which refer to your particular paper, and write a brief summary of what you find. Your summaries should be written for the benefit of other statisticians who have not read the papers. Any mathematical symbols which you use should be defined, unless standard such as $N(\mu, \sigma^2)$.

Bear in mind that you are unlikely to understand every word of the papers, and this is in any case unnecessary for getting the general 'message' of the papers. You should also realize that papers do sometimes get published which contain errors and/or views which are controversial to say the least.

Andrews, D. F. (1972) Plots of high-dimensional data. *Biometrics*, **28**, 125–36.

Broadbent, S. (1980) Simulating the ley hunter. *J. R. Stat. Soc., Series A*, **143**, 109–40.

Chatfield, C. (1978) The Holt–Winters forecasting procedure. *Appl. Stat.*, **27**, 264–79.

Chernoff, H. (1973) The use of faces to represent points in k-dimensional space graphically. *J. Am. Stat. Assoc.*, **68**, 361–8.

Oldham, P. D. and Newell, D. J. (1977) Fluoridation of water supplies and cancer – a possible association? *Appl. Stat.*, **26**, 125–35.

Preece, D. A. (1981) Distributions of final digits in data. *The Statistician*, **30**, 31–60.

(*Note*: Teachers can readily supplement the lists in both (a) and (b) so as to give all students in a class a different subject. See also Hawkes (1980) for some further suggestions for (b).)

Exercise G.7   Final miscellanea

(a) *Quality control*   A company which manufactures packets and boxes of rubber bands has called you in to discuss quality control, because of increasing complaints from purchasers. These complaints about product quality include 'too many bands are broken' and 'some packets are nearly empty'. Discuss the questions which need to be answered and the problems which are likely to arise in implementing a new quality control scheme.

(b) *Understanding the χ^2-test*   Given a two-way table of frequencies, the χ^2-test statistic, for testing that rows and columns are independent, is clearly non-negative. If all the observed frequencies are identical, then the expected frequencies will also be identical and the χ^2 statistic will be zero. Construct a (2×2) table of frequencies, which are all **unequal**, for which the χ^2-statistic is also exactly zero. If you really understand the χ^2-test, you will find this easy.

Discussion of Exercise G.1

The argument is completely nonsensical, and you will be pleased to learn that the verdict was overturned by the California Supreme Court. Four reasons were given for overturning the verdict. First no empirical evidence was given to support the suggested probability values, such as prob. (partly yellow car) = $1/10$. Second, the use of the product rule assumes independence. This is plainly false. For example, 'growing a beard' and 'growing a moustache' are positively correlated. Third, the probability calculations do not take account of the possibility of witnesses being mistaken or lying, or of the criminals wearing disguise. Fourth, even if the probability calculations were correct, they give an (unconditional) probability that a 'random' couple will have all the characteristics. What is required is the (conditional) probability that a couple having all these characteristics is innocent. If there are say 24 million couples in the suspect population, then we expect to find two couples with all the characteristics so that the conditional probability that either is innocent is $1/2$, not $1/12\,000\,000$.

A similar catastrophe is discussed by Huff (1959, p. 113). Regrettably the product law is widely misused. In addition, there are many other examples which demonstrate the importance of computing a conditional probability rather than an unconditional probability, and moreover of conditioning on the right event, I recall discussing the case of a woman who had given birth to five girls and wanted to know the probability of having a boy next time. As the probability of having six girls in a row is only $1/2^6$, it was falsely argued that the chance of having a boy must be $(1 - 1/2^6) = 0.985$! The conditional probability, based on empirical evidence, turns out to be very close to $1/2$, which is of course also the probability assuming equally likely outcomes. Conditional on what, I leave to the reader.

One of the earliest uses of probability in law was in the notorious Dreyfus case in France in 1899, when samples of handwriting had to be compared. When the case was reviewed several years later, it became apparent that the mathematical arguments which had been used were in fact false, but that none of the lawyers involved had understood them at the time. Nevertheless, the judges allowed themselves to be impressed by the 'scientific' nature of the evidence.

Statistics and probability may well have an increasingly important role to play in courts of law. However, it is imperative that they be used both correctly and intelligibly, in order to prevent statistics from being discredited in the public eye. An interesting review of the difficulties of introducing statistics into courts of law is given by Meier (1986).

Notes on Exercise G.2

There are two aspects to this exercise: first, carrying out the analysis, which may be regarded as reasonably straightforward; second, having the ability to communicate the results to a non-statistician. Students typically get little help in the latter important skill.

We look at the analysis first. The request to 'analyse the data' is typically vague. The first thought which occurs to you may be that arrivals are completely random (i.e. form a Poisson process). However, the exact time of admission is unknown, although we know that most admissions are at planned times during the day. In any case a cursory examination of the data shows that admissions are not equally likely on different days of the week. In particular there are few arrivals on Sundays, as might be expected.

The two main sources of variation in daily admission rates are likely to be day-of-the-week variation and time-of-the-year variation, and they each need to be assessed and described. For day-of-the-week variation, the frequency distribution of admissions on different days of the week should be constructed. Then a χ^2 goodness-of-fit test could be carried out to test the null hypothesis that arrivals are equally likely on different days of the week. In order to get equal expected frequencies and simplify the analysis, it is helpful to 'drop' the last two days of the year (it is a leap year) so that there are exactly 52 weeks. The device of omitting a small part of the data to simplify the analysis (or even improve it – see Example F.2) is widely used in practice but rather neglected in textbooks. The total number of arrivals through the 52 weeks is 878, but only 45 arrive on Sunday. The latter compares with an expected value of $878/7 = 125.4$ and contributes $(45 - 125.4)^2/125.4 = 51.4$ to the χ^2-statistic. Even by itself this contribution is so large, compared with the degrees of freedom $(7 - 1 = 6)$, that the overall test statistic is highly significant, leading to rejection of the null hypothesis. In such a situation it is unnecessary to calculate all of the χ^2 statistic and this is another widely used short cut which is rarely mentioned in textbooks. Admissions are clearly less likely on Sundays. Indeed, the reduced number of arrivals on Sunday is so obvious that it may be more fruitful to apply the χ^2 test to the remaining six days.

Turning to time-of-the-year variation, the 'obvious' analysis is to find the numbers of admissions in different months. However, different months are of different lengths and contain different numbers of Sundays. It is much easier to split the year into 13 four-week periods giving the frequency distribution shown in Table G.2. Under the null hypothesis of a constant arrival rate, the χ^2-statistic of 9.4 is less than its degrees of freedom, namely $(13 - 1) = 12$, and so is clearly not significant. Thus there is no evidence of variation through the year.

Table G.2 Observed and expected number of arrivals at a psychiatric hospital in successive four-week periods

Weeks	Observed	Expected*
1–4	74	67.5
5–8	64	67.5
9–12	54	67.5
13–16	72	67.5
17–20	62	67.5
21–24	72	67.5
25–28	67	67.5
29–32	76	67.5
33–36	73	67.5
37–40	66	67.5
41–44	58	67.5
45–48	78	67.5
49–52	62	67.5

χ^2 *test*

$$\chi^2 = \sum (\text{obs.} - \text{exp.})^2/\text{exp.}$$
$$= 637.2/67.5$$
$$= 9.4$$

$$\text{c.f. } \chi^2_{12,0.05} = 21.03$$

* Each expected frequency $= 878/13$.

Finally, we consider the question as to whether there is any cyclic behaviour other than the regular within-week variation. The latter makes it more difficult to spot longer-term cyclic behaviour in daily admission rates. We could apply correction factors to daily rates, but it is probably easier to carry out a time-series analysis of weekly totals. Using MINITAB or some similar package, the autocorrelation function of successive weekly totals may be calculated to see if successive values are correlated. It turns out that all the coefficients are 'small'. For example, the first-order autocorrelation coefficient is only 0.01. With a series length 52 (weeks), the modulus of an autocorrelation coefficient must exceed $2/\sqrt{52} = 0.28$ to be significant at the 5% level, although, when testing a large number of coefficients, the level of significance needs considerable adjustment to avoid spurious significant results. In this case only the coefficient at lag 6, namely -0.26, is nearly significant. The lack of structure in the autocorrelation function means that there is no evidence of cyclic behaviour.

The reader may like to fit a Poisson distribution to the frequency distribution of weekly totals and show that a very good fit arises (and I emphasize that these are real data).

The results of the analysis now need to be described in a clear common-sense way to someone who will probably not know what is meant by 'Poisson process', 'χ^2-test', and (especially) 'autocorrelation'. This is excellent practice, and I suggest you read the general advice given in Chapter 11 of Part One. There is little point in my attempting a definitive description as it is you who needs the practice! However, I offer a few tips. You should always make a special effort to make all graphs and tables clear and self-explanatory. Here you want to present the frequency distribution of admission numbers through the week and through the year. Whether they are best presented in a table (as in Table G.2) or in a bar chart is debatable. Try both. As to cyclic behaviour, it is probably best simply to say that a time-series analysis revealed no evidence thereof, rather than to try to explain the mechanics involved. Summarize your conclusions briefly, particularly the form of the within-week variation. When you think you have finished, read through your letter one last time and try to imagine that you are the (non-statistical) recipient of the letter.

As well as summarizing the data, you may also wish to raise additional queries. For example, how much notice do psychiatrists take of the number of empty beds when deciding to admit patients?

The data set given here is somewhat similar to Example A of Cox and Snell (1981), which considers arrival times at an intensive care unit and provides a detailed analysis. There the exact time of arrival is recorded so that time-of-day variation may also be investigated.

Two morals

1. When describing the variation in a set of data, the likely sources of variation should be listed and then assessed.
2. Being a good analyst is only part of the statistician's task. It is equally important to be able to communicate the results effectively.

Notes on Exercise G.3

What is meant by a sequence of 'random numbers' as applied to a sequence of digits which may take any integer value from 0 to 9? The sequence is said to be random if all 10 possible integers are equally likely to occur, and if different outcomes are mutually independent.

The problem of testing random numbers has a long history and has many applications, particularly now that random digits are generated by computers for a variety of purposes, particularly for **simulation** (e.g. Morgan, 1984). Simulation is used to solve problems which are difficult or impossible to solve analytically by copying the behaviour of the system under study by generating appropriate random variables.

It must be appreciated that it is impossible to test all aspects of randomness with a single overall test. Rather there are numerous tests for examining different aspects of randomness (Morgan, 1984, Chapter 6). There is an added complication here in that the so-called random digits come in groups of size six.

The first 'obvious' property to test is that all 10 possible digits are equally likely. Count the number of zeros, the number of ones, and so on, either in the whole data set or in part of it. Then compare the observed frequencies with the expected frequencies, namely $n/10$ where n is the number of digits examined. A χ^2 goodness-of-fit test on nine degrees of freedom is appropriate.

We must also check that the order of digits is random. At its simplest, this means checking adjacent pairs of digits. Consider non-overlapping pairs of digits (to ensure independence between pairs) and compute the (10×10) matrix containing the observed frequencies with which digit j is followed by digit k. With m non-overlapping pairs, the expected frequencies under randomness are $m/100$ and a χ^2-test on 99 DF is appropriate. A difficulty here is that m needs to exceed 500 in order to get expected frequencies exceeding five so that a χ^2 approximation is valid. Otherwise cells will need to be grouped.

As well as ordering, we may also be interested in other features of the data. In particular, we may wish to examine some special properties of the groups of six numbers. For example, we may suspect a tendency for digits to be repeated, and can easily devise a suitable test. If the digits are really random, the probability that a six-digit number contains at least two consecutive identical digits is $(1 - 0.9^5) = 0.410$ and this can be compared with the observed proportion of groups having this property.

One problem with trying several different tests on the same data is that a significant result is likely to occur eventually even if the data really are random. It is therefore wise to adjust the required level of significance or repeat a test on a new set of numbers. For the data in Table G.3, I see no reason to question the randomness hypothesis.

Solution to Exercise G.4

The reader may be tempted to treat the data as a 2×2 contingency table and carry out a χ^2-test to see if rows and columns are independent. The result is significant and so it does appear at first sight that women are treated more leniently than men. It is easy to speculate why this should be so. For example, the courts may be sympathetic to women who have young children to look after or who have acted under male coercion. However, there is more in this exercise than meets the eye. Whenever two variables appear to be associated it is always advisable to ask if this association can be explained by a common association with a third variable.

In this case, it is worth considering 'number of previous convictions' as a third variable, and this reveals the surprising result that women are actually treated **less** leniently than men. A three-way table of frequencies, categorized by sex, severity of sentence and number of previous convictions, was constructed and is shown in Table G.4. Women generally have fewer previous convictions

Table G.4 Three-way table of frequencies

	Male		Female	
No. of previous convictions	Sample size	No. of severe sentences	Sample size	No. of severe sentences
0	10	0	70	20
1–2	20	8	20	10
≥ 3	70	52	10	10
Overall	100	60	100	40

but it can be seen that they are treated less leniently than men, in each row of the table, even though they appear to fare better overall. This curious effect is related to the so-called **Simpson's paradox**, and you may have to look at the table for some time to convince yourself that you are not 'seeing things'. Similar examples are given by Hooke (1983, Chapter 13) and by Hand (1994, Example 5). Hand's (1994) Example 3 is another interesting case where it is difficult to believe what one sees.

Table G.4 demonstrates that collapsing data onto two dimensions can have a misleading effect. This is something to watch out for when considering two-way tables or scatter diagrams obtained from multivariate data. (**Note:** The data are partially fictitious, as they have been adapted from real data to make the point more forcibly.)

Moral

An association between two variables does not prove causation, and in particular may sometimes be explained by a common association with a third variable.

Notes on Exercise G.5

(a) You should start by querying the form of the data. Why have the given dose levels been chosen and why are they not equally spaced? What determines the dose level for a particular patient and in what units is it measured? Why are the sample sizes unequal? Will the use of unrandomized case histories bias the results?

In the absence of clear answers and accepting the data (with caution) as they stand, the 'obvious' first step is to plot the proportions with after-effects (r_i/n_i) against dose level on ordinary graph paper. The proportion appears to rise with dose level. Fitting a smooth curve by eye, we would guess that the ED_{50} is 'near' 2.0. The fitted curve is probably S-shaped (as proportions are constrained to lie between 0 and 1) and this is hard to fit, especially as the sample sizes vary so that the points are not equally important. If, instead, the data are plotted on normal probability paper, then we would expect the relationship to be roughly linear. This graph 'stretches' the proportion scale in a non-linear way. Fitting a straight line by eye, the ED_{50} is between about 1.9 and 2.0.

This crude analysis may be adequate for some purposes, but if you have studied logit analysis (Appendix A.9) then the GLIM package can be used (Example B.4 in Appendix B) to get a model relating the probability of getting after-effects, say p, to dose level, x. We cannot fit a linear model as p is constrained to lie between 0 and 1. Instead we fit logit $(p) = \log[p/(1 - p)]$ as a linear function of x, assuming that the number of patients with after-effects has a binomial distribution. This is then a generalized linear model. We find

$$\text{logit } (p) = -1.52 + 0.781x$$

A probit analysis could also be carried out and this gives

$$\text{probit } (p) = -0.931 + 0.477x$$

The two fitted models look quite different but actually give very similar results over a wide range of x-values. In particular, both have a positive gradient with x and give ED_{50} values equal to 1.95. The standardized residuals at each dose level may be examined to check that there is no evidence of model inadequacy.

There are various benefits of model fitting here. In particular, the results take account of the different sample sizes at different x-values, and it is possible to get a confidence interval for the ED_{50}.

We may summarize the results by saying that the chance of after-effects increases from about 30% at the lowest dose level of 0.9 to over 80% at the highest dose level of 4.0, with an ED_{50} at $x = 1.95$.

Moral

This exercise provides a simple example of the benefits which may arise from fitting a generalized linear model.

(b) At first sight, the data are of similar form to that in part (a). However, in important respects, they are quite different. In particular, the explanatory variable as tabulated is ordinal rather than quantitative, which disallows a standard logit analysis unless one can get the exact dose levels. Moreover, even a cursory examination of the data reveals that the proportion of rats dying is zero or near zero at dose levels 1, 2, and 3 but jumps to one at levels 4 and 5. The LD_{50} dose level is therefore clearly somewhere between dose levels 3 and 4. It also seems intuitively clear that the data do not allow us to say any more as the dose levels are too far apart to give any information as to what happens between these levels 3 and 4.

Of course one would expect the probability of a rat being killed to increase from zero to one as the dose increases. If the exact dose levels were known, the usual model which would be fitted to such data is a logit model with dose level as the explanatory variable. Textbook examples typically show the sample proportions dying (or surviving or suffering after-effects or whatever is being measured) going from, say, 0.1 to 0.9. However, with the widely separated dose levels used in this screening experiment, the sample proportions jump suddenly from nearly zero to nearly one. The technical literature might well expend (waste?) energy debating whether existing algorithms for fitting logit models will converge for such data. Perhaps a more important question is whether it is really reasonable to expect or want the algorithms to 'work' here, or whether it would be better for an algorithm to stop and indicate a problem with the data. In my view, all that can sensibly be said here is that the proportion of rats dying increases from near zero to near one as the dose level is increased from level 3 to level 4, and that the LD_{50} must lie somewhere in between.

Moral

If the data are inadequate to answer a given question, say so!

Notes on Exercise G.6

(a) Sources of statistics are briefly discussed in Chapter 9 of Part One. You should have relatively little trouble in finding most of these statistics.

(b) Not all journals are taken by all libraries, so you may have trouble locating a particular journal. If so, try a different reference or ask for a photocopy using the inter-library loan system.

Bear in mind that the reader wants an overview of the paper and the subsequent references. Do not get bogged down in mathematical detail. On the other hand, you should carefully explain any important mathematical concepts in the given paper. Remember to say what you think of the paper. Is it clearly written? Do you agree with it?

A critical summary should review and evaluate a paper so that a prospective reader can assess if it is worth reading in detail. The objectives of the paper should be stated together with an assessment of how well the author has succeeded. The summary might also explain the methods used and indicate any underlying assumptions and any limitations (see also Anderson and Loynes, 1987, section 4.6.3).

Brief notes on Exercise G.7

(a) When this problem came to me, I was relatively inexperienced and diffident about giving advice. Nevertheless, I asked to be shown around the factory to see how the rubber bands were made, and then asked to inspect their quality control procedure. It turned out that they had no quality control whatsoever. Anything I suggested would be an improvement! Having found out how the factory operated, and what was feasible statistically (there was no statistical expertise on hand at all), I recommended a very simple plan. Samples of packets and boxes were weighed at regular intervals and the weights plotted on a simple chart. Warning and action lines were inserted. The contents of sample bags were also inspected at regular intervals, but less frequently. It takes a long time to count the number of complete rubber bands in a packet size 500.

(b) There is only one degree of freedom when testing a 2×2 table. This suggests that we can make three of the four frequencies anything we like. Let us assume the table is, say

4	12
10	?

To make the χ^2-statistic zero, we simply take the missing frequency to be the same multiple of 10, as 12 is of 4, namely $12 \times 10/4 = 30$. The null hypothesis of row-and-column independence implies that *ratios* of expected frequencies are the same in each pair of rows (and each pair of columns). This is why linear models are fitted to the logarithms of expected frequencies in contingency tables, giving rise to what are called log-linear models (Appendix A.9).

H
Collecting data

This chapter contains three exercises which cover some aspects of data collection. Some readers may wish to study this section before some of the earlier exercises, as statistical analysis depends on having reliable, representative data. However, some aspects of data collection are appreciated more after data analysis has been studied. Thus this section is placed at the end of Part Two. (Which comes first – the chicken or the egg?)

Some general remarks on data collection are made in Chapter 4 of Part One and in sections A.10, A.11 and A.12 of Appendix A.

Because of the time-consuming nature of data-collection exercises, this section is shorter than might be expected from its relative importance. It is certainly true that the main way to appreciate the difficulties of data collection is by doing it, and it is to be hoped that students have carried out some experiments to collect data in earlier courses (e.g. Scott, 1976), even if they were only very simple experiments like coin tossing. Experiments help to stimulate student interest, and help develop statistical intuition.

Exercise H.1 Sample surveys

Suppose you want to carry out a survey to investigate one of the following topics. In each case carefully identify the target population, design a suitable questionnaire, and discuss how you would get a representative sample from the given target population.

(a) The financial problems of students at a particular college or university.
(b) The popularity of the government of your country among adults.
(c) The views of the 'general public' on capital punishment, abortion, tax evasion or some similar controversial ethical topic. How could you get the views of a more specialized population such as doctors or policemen?
(d) Agricultural production of a particular crop in a particular region.

Exercise H.2 Designing a comparative experiment

This is an artificial exercise on constructing experiments to compare a set of treatments. Suppose there are t treatments and m experimental units and that one (and only one) treatment can be applied to each unit. Further, suppose that the experimental units can be divided into b reasonably homogeneous blocks of size k, such that $m = bk$. Construct a suitable design if:

(a) $t = 4$, $m = 12$, $b = 3$, $k = 4$.
(b) $t = 4$, $m = 12$, $b = 4$, $k = 3$.
(c) $t = 4$, $m = 12$, $b = 6$, $k = 2$.
(d) $t = 5$, $m = 15$, $b = 5$, $k = 3$.

For convenience denote the treatments by A, B, C, D, ... and the blocks by I, II, ...

For design (a), describe how you could use a table of random digits to allocate the treatments randomly to the experimental units.

Exercise H.3 A study on longevity

In a study on longevity, a doctor decided to compare the age at death of marriage partners to see if the pairs of ages were correlated. He questioned 457 consecutive patients who were over 40 years old and had come to consult him for some medical reason. Each patient was asked if his/her parents were alive or dead, and, if the latter, how old they were when they died. After omitting patients who still had one or both parents alive, the doctor found a positive correlation between the ages of dead parents. Comment on this conclusion and on the way the data were collected.

Notes on Exercise H.1

These notes concentrate on problem (a). The obvious target population is 'all students at the particular college', but this is not as precise as it appears. It is probably sensible to exclude part-time students (who may have special problems and be difficult to contact) and postgraduate students. Students who are temporarily absent from college (e.g. on a spell of industrial placement) should also be excluded. Can you think of any other possible exclusions?

To get a representative sample, we must not interview students haphazardly on the campus. Rather a quota or random sample must be chosen. For a random sample, we need a list of students (the sampling frame), from which a random sample may be selected using random numbers. However, a simple random sample would be too difficult and expensive to take. It is much better to use a form of stratified, multi-stage random sampling. Divide the departments of the college (e.g. physics, history) into strata, which could be 'sciences', 'arts' and 'technologies'. Take a random sample (perhaps size one) of departments from each stratum. Obtain lists of students for each year (or from randomly selected years) in the selected departments, and choose a random sample from each year. This will produce some clustering which may lead to correlated responses, but also means that sampling is inexpensive so that a larger sample may be taken for the same cost. Alternatively, students may be approached at random on the campus until certain quotas have been satisfied. For example, a representative proportion of men and women is needed, as well as a representative proportion of first-year students, science students and so on.

Designing the questionnaire is not easy and will take longer than most students expect. It should start with questions on basic demographic details (name, department, year, sex, etc.). Then the student's income needs to be assessed as well as any fixed expenditure. It is probably unwise to ask questions such as 'How much do you spend on entertainment per week?' as this is non-essential expenditure which varies considerably from week to week and is difficult to estimate. Avoid open questions such as 'How would you describe your overall financial situation?', as the answers may be difficult to code and analyse unless a clearly specified list of alternatives is given, such as (a) more than adequate, (b) just about adequate, (c) rather inadequate, (d) grossly inadequate. Wherever possible, questions should have a clearly defined answer (e.g. 'What finance, if any, do you receive from the government?'). Avoid leading questions such as 'Would you agree that students do not receive enough financial help from the government?' Allow for the possibility of 'Don't know' or 'Not applicable' in the suggested answers to some questions.

A small pilot survey is essential to try out the questionnaire. You may be amazed at what can go wrong when this is not done. When the author tried out his first attempt at a questionnaire, he found that some students did not know the size of the grant they received, or what their parents contributed. Students who paid their rent monthly found difficulty in assessing the corresponding weekly rate. And so on.

For problem (b), the target population is apparently all adults who are resident in the country who are eligible to vote. Can you think of any problems with this definition? Neutral questions are essential. Then the country should be divided into reasonably homogeneous strata, so that smaller areas may be randomly selected from each stratum and then further subdivided if necessary. Eventually a random selection of adults can be taken from the electoral register. Alternatively, a quota sample could be taken by selecting quota samples in different representative areas of the country (not an easy task!).

For problem (c), it is very difficult to compose neutral questions. You should also bear in mind that some respondents may be unwilling to cooperate or may not give their true opinions.

For problem (d), a major difficulty is that of getting a sampling frame.

Notes on Exercise H.2

(a) When the block size is the same as the number of treatments, it is sensible to make one observation on each treatment in each block. If the treatments are applied randomly, then we have what is called a **randomized block design**.

If block I contains four experimental units, denoted by g, h, i, j, then the four treatments A, B, C, D may be allocated to them, using random digits in many different ways to ensure fair allocation. One crude method is to ignore digits 4 to 9, and to allocate A to g, h, i or j according as to whether the first usable digit is 0, 1, 2 or 3. And so on. A more efficient method can easily be devised.

(b) When the block size is less than the number of treatments, we have an **incomplete block design**. When $b = t$ and $k = t - 1$, the 'obvious' way to proceed is to omit each treatment from just one of the blocks. This leads to a balanced incomplete block design since each pair of treatments occurs within the same block the same number of times. Thus we could have: block I – ABC; block II – ABD; block III – ACD; block IV – BCD. Randomize the allocation within blocks.

(c) A balanced incomplete block design is also possible here.

(d) It is not possible to construct a balanced design. Use your common sense to construct an incomplete block design with as much symmetry as possible. Alternatively, consult a textbook (e.g. Cochran and Cox, 1957) which gives partially balanced designs. It only took me a few minutes to construct the following design by trial and error so that each treatment occurs three times and so that each pair of treatments occurs within the same block either once or twice: block I – ABE:block II – CDE:block III – ABD:block IV – BCE:block V – ACD. Advances in computing software means that balance, although still desirable, is not as important as it used to be.

Notes on Exercise H.3

It is important to realize that the sample collected here is not a random sample from the total population of all married couples. The patients are all over 40 years old, they are all ill (which is why they are seeing a doctor), and they all have both parents dead. Thus the sample is biased in three different ways. This means that the results should not be taken as representative of the population as a whole.

Many results which are reported in the media or in the published literature are based on biased samples, although the bias is not always as obvious as it is here. Another common form of biased sample is that formed by people who write letters to newspapers, to radio programmes or to elected representatives. These people feel strongly about a particular topic and are self-selecting. Any survey which produces a low response rate may also give biased results because the people who respond may be different to those who do not respond.

Moral

Be on the lookout for biased samples.

PART THREE
Appendices

Appendix A
A digest of statistical techniques

This appendix is a concise reference source containing brief notes on a variety of statistical topics. These topics are a personal selection and are not intended to be comprehensive. Important definitions and formulae are given together with some brief advice and warnings. Key references are given for further reading.

This appendix is intended as an *aide-mémoire* for students and researchers who have already studied the topics. It is not suitable for learning topics and is not intended for use as a 'cookbook' by statistical novices. It will need to be supplemented by other textbooks for further details.

Standard abbreviations

pdf   probability density function
cdf   cumulative distribution function
CI   confidence interval
SS   sum of squares
DF   degrees of freedom
MS   mean square
ANOVA   analysis of variance

General references

There are many good introductory textbooks on statistics, at varying degrees of mathematical difficulty, which cover basic statistical methods. They include:

Box, G. E. P., Hunter, W. G., and Hunter, J. S. (1978) *Statistics for Experimenters*, Wiley, New York.
Chatfield, C. (1983) *Statistics for Technology*, 3rd edn, Chapman and Hall, London.
Moore, D. S. and McCabe, G. P. (1993). *Introduction to the Practice of Statistics*, 2nd edn, W. H. Freeman, New York.

Snedecor, G. W. and Cochran, W. G. (1980) *Statistical Methods*, 7th edn, Iowa State University Press, Ames, Iowa.

A good introduction to more theoretical topics is:

Bain, L. J. and Engelhardt, M. (1987) *Introduction to Probability and Mathematical Statistics*, 2nd edn, PWS-Kent, Boston.

There are many more advanced books on a variety of topics, some of which are referred to in the appropriate section below. In addition, it may sometimes help to consult a dictionary or encyclopedia such as:

Kotz, S. and Johnson, N. L. (eds) (1982–) *Encyclopedia of Statistical Sciences* (in 9 volumes), Wiley, New York.

Kruskal, W. H. and Tanur, J. M. (eds) (1978) *International Encyclopedia of Statistics*, (in 2 volumes), Collier-Macmillan, New York.

Marriott, F. H. C. (1990) *A Dictionary of Statistical Terms*, 5th edn, Longman, Harlow.

Tietjen, G. L. (1986) *A Topical Dictionary of Statistics*, Chapman and Hall, New York.

A.1 Descriptive statistics

The calculation of summary statistics and the construction of graphs and tables are a vital part of the initial examination of data (IDA). These topics are discussed in section 6.5 of Part One and in the exercises of Part Two (especially Chapter A). The important advice given there will not be repeated here.

Given a sample of n observations, say $x_1, \ldots, x_n$, the three main measures of **location** are the (arithmetic) **mean**, $\bar{x} = \Sigma x_i/n$; the **median**, which is the middle value of $\{x_i\}$ when they are arranged in ascending order of magnitude (or the average of the middle two observations if n is even); and the **mode**, which is the value which occurs with the greatest frequency. An alternative measure of location is the **trimmed mean**, where some of the largest and smallest observations are removed (or trimmed) before calculating the mean. For example, the MINITAB package routinely calculates a 5% trimmed mean where the smallest 5% and largest 5% of the values are removed and the remaining 90% are averaged. This gives a robust measure which is not affected by a few extreme outliers.

The two common measures of **spread**, or variability, are:

1. **standard deviation** $= s = \sqrt{[\Sigma(x_i - \bar{x})^2/(n - 1)]}$

2. **range** $=$ largest $(x_i) -$ smallest (x_i).

The range is easier to understand than the standard deviation but tends to increase with the sample size, n, in roughly the following way for normal data:

$$s \simeq \text{range}/\sqrt{n} \qquad \text{for } n < \text{about } 12$$
$$s \simeq \text{range}/4 \qquad \text{for } 20 < n < 40$$
$$s \simeq \text{range}/5 \qquad \text{for } n \text{ about } 100$$
$$s \simeq \text{range}/6 \qquad \text{for } n \text{ above } 400.$$

These rough guidelines can be helpful in using the range to check that a sample standard deviation has been calculated 'about right'. Despite the dependence on sample size, the range can also be useful for comparing variability in samples of roughly equal size.

An alternative robust measure of spread is the **interquartile** range given by $Q_3 - Q_1$, where Q_1, Q_3 are the lower (first) and upper (third) quartiles, respectively. Thus, Q_1 is the value below which there are a quarter of the observations. The median can be regarded as the second quartile. Interpolation formulae may be needed to calculate Q_1 and Q_3. The quartiles are special types of percentile – a value which cuts off a specified percentage of the distribution.

Tukey (1977) has introduced several descriptive terms which are growing in popularity. **Hinges** are very similar to the (upper and lower) quartiles, while the **H-spread** is similar to the interquartile range. A **step** is 1.5 times the H-spread. An **inner fence** is one step beyond the hinges, while an **outer fence** is two steps beyond the hinges. Observations outside the outer fences can be regarded as extreme **outliers**.

Let $m_k = \Sigma(x_i - \bar{x})^k/n$ denote the kth moment about the mean. A coefficient of **skewness**, which measures the lack of symmetry, is given by $m_3/m_2^{3/2}$. A coefficient of **kurtosis** is given by $[(m_4/m_2^2) - 3]$. This measures whether the observed distribution is too peaked or too heavy-tailed as compared with a normal distribution. Both these shape coefficients should be close to zero for normal data.

Various types of graph are discussed in section 6.5.3 of Part One. Most will be familiar to the reader, but it seems advisable to include further details here on **probability plotting**. The general idea is to rank the data in ascending order of magnitude and plot them in such a way as to show up the underlying distribution.

Given a sample size n, denote the ordered data (i.e. the order statistics) by $x_{(1)}, x_{(2)}, \ldots x_{(n)}$. Traditional probability plots are obtained by plotting the ranked data against the sample (or empirical) cumulative distribution function (cdf) on special graph paper called probability paper. The proportion of observations less than or equal to $x_{(i)}$ is i/n, but this can provide a poor estimate of the underlying cdf. (For example, it implies that the chance of getting a value greater than $x_{(n)}$ is zero.) Three common estimates of the cdf at $x_{(i)}$ are $i/(n + 1)$, $(i - \frac{1}{2})/n$ and $(i - \frac{3}{8})/(n + \frac{1}{4})$, and there is little to choose between them. If we simply plot, say, $i/(n + 1)$ against $x_{(i)}$ on graph paper with linear scales, we typically get an S-shaped curve called the sample cdf. (If the two axes are transposed we have what is called a **quantile plot**. The pth quantile, $Q(p)$, (or the $100p$th percentile) of a distribution is the value below which a proportion p of observations from that distribution will lie.) This graph may be helpful but the information is easier to interpret if plotted on special graph paper constructed so that the sample cdf is approximately a straight line if the data come from the particular distribution for which the graph paper is constructed. For example, normal probability paper has one scale linear while the other scale is chosen in a non-linear way so as to transform the normal cdf to a straight line. Weibull paper is also widely used in reliability work. Note that percentiles can be easily estimated from plots of this type.

Computers generally construct probability plots in a rather different way by plotting the ordered observations against the expected values of the order statistics for a random sample of the same size from the specified distribution of interest (often normal). Expected order statistics can be found for a particular distribution by an appropriate calculation or approximation. For the normal distribution, for example, MINITAB calculates what it calls **normal scores** (sometimes called **rankits**) by calculating $\Phi^{-1}[(i - \frac{3}{8})/(n + \frac{1}{4})]$, where Φ denotes the cdf of the standard normal distribution. Thus computer probability plots are likely to have two linear scales. Such plots are sometimes called **theoretical quantile-quantile plots** as they essentially plot the observed quantiles (the ranked data) against theoretical quantiles. Departures from linearity in a probability plot suggest the specified distribution is inappropriate, but if the departures cannot easily be spotted by eye, then they are probably not worth worrying about. For small samples (e.g. $n <$ about 20), linearity can be hard to assess and it can be helpful to simulate plots from the specified distribution to see what sort of variability to expect.

A.2 Probability

A sound grasp of basic probability theory and simple probability models is needed both to understand random events and as a basis for statistical inference. Probability and statistics are complementary but inverse subjects in that statistics is concerned with **inductive** inference (DATA → MODEL), while probability is concerned with **deductive** inference (MODEL → behaviour of system).

There are some philosophical problems in deciding what is meant by 'probability', and in establishing a set of sensible axioms to manipulate probabilities. Different types of probability include equally-likely probabilities, long-run frequentist probabilities, and subjective probabilities (see also section 7.4), but they will not be discussed here.

Suppose you are interested in finding the probabilities of the different outcomes of one or more experiments or trials. Begin by finding the set of all possible outcomes of the experiment, called the **sample space**. An **event** is a subset of the sample space. If E_1 and E_2 denote two events, and $P(E)$ denotes the probability of event E, then the two most important rules for manipulating probabilities are:

$$P(E_1 \cup E_2) = P(E_1) + P(E_2) - P(E_1 \cap E_2) \qquad \text{(A.2.1)}$$

the **general addition law**, and

$$P(E_1 \cap E_2) = P(E_1)P(E_2 | E_1) \qquad \text{(A.2.2)}$$

the **general multiplication law**.

If the two events are **mutually exclusive**, then $P(E_1 \cap E_2) = 0$ and equation (A.2.1) simplifies to

$$P(E_1 \cup E_2) = P(E_1) + P(E_2) \qquad \text{(A.2.3)}$$

the addition law for mutually exclusive events.

If E_1 and E_2 are **independent**, then the conditional probability $P(E_2 | E_1)$ is the same as the unconditional probability $P(E_2)$ and (A.2.2) reduces to

$$P(E_1 \cap E_2) = P(E_1)P(E_2) \qquad \text{(A.2.4)}$$

the product law for independent events. Sadly this rule is often used when events are not independent (Exercise G.1).

There are many good introductions to probability such as Bain and Engelhardt (1987). The classic text by Feller (1968) is for the more advanced reader.

A.3 Probability distributions

A random variable, X, takes numerical values according to the outcome of an experiment. A random variable may be discrete or continuous according to the set of possible values it can take. A **discrete** distribution is usually defined by a point probability function, $P(X = r)$ or $P(r)$, while a **continuous** distribution may be described by a **probability density function** (pdf), $f(x)$, or equivalently by a **cumulative distribution function** (cdf), $F(x)$, such that

$$F(x) = \text{Prob}(X \leq x)$$

$$= \int_{-\infty}^{x} f(u) \, du.$$

The inverse relationship is $f(x) = dF(x)/dx$.

The mean (or expected value or expectation) of a random variable, X, is given by

$$E(X) = \begin{cases} \sum rP(r) & \text{in the discrete case} \\ \int_{-\infty}^{\infty} xf(x) \, dx & \text{in the continuous case} \end{cases}$$

The expectation operator can be defined more generally by

$$E[g(X)] = \begin{cases} \sum g(r)P(r) & \text{in the discrete case} \\ \int g(x)f(x)\,dx & \text{in the continuous case} \end{cases}$$

where g denotes a function. In particular, the variance of a probability distribution is given by $E[(X - \mu)^2]$ where $\mu = E(X)$. There are numerous rules for manipulating expectations, such as $E(X_1 + X_2) = E(X_1) + E(X_2)$ which applies to any two random variables, X_1 and X_2 (with finite means). The properties of some common distributions are listed in Table A.3.1. Note that the **probability generating function** of a (non-negative) discrete distribution is defined by $\sum_{r=0}^{\infty} P(X = r)s^r$.

If the random variable X has a **normal distribution** with mean μ and variance σ^2, then we write $X \sim N(\mu, \sigma^2)$, where $\sim$ means 'is distributed as'. The standard normal distribution arises when $\mu = 0$, $\sigma = 1$. Some useful results are as follows:

1. If $X \sim N(0, 1)$, then $Y = X^2$ is said to have a **chi-squared** (χ^2) distribution with one degree of freedom (DF). If $X_1, \ldots X_n$ are independent $N(0, 1)$ variables, then

$$\sum_{i=1}^{n} X_i^2 \sim \chi_n^2.$$

2. If Y has a **gamma** distribution with parameters r and $\lambda = 1$, then $2Y$ has a χ^2 distribution with $2r$ DF or χ_{2r}^2, for short.
3. If $X \sim N(0, 1)$ and $Y \sim \chi_v^2$, and X and Y are independent, then the random variable

$$t = \frac{X}{\sqrt{(Y/v)}}$$

is said to have a **t-distribution** with v DF.
4. If X_1, X_2 are independent χ^2 random variables with v_1, v_2 DF respectively, then the random variable

$$F = \frac{X_1/v_1}{X_2/v_2}$$

is said to have an **F-distribution** on (v_1, v_2) DF.
5. A variable, Y, has a **lognormal** distribution if $X = \ln Y \sim N(\mu, \sigma^2)$. Then $E(Y) = \exp(\mu + \sigma^2/2)$.
6. The **exponential** distribution is a special case of both the gamma distribution (with $r = 1$) and the **Weibull** distribution (with $m = 1$, see Table A.3.1).
7. The **Erlang** distribution is a special case of the gamma distribution with r a positive integer.

Table A.3.1 The properties of some common distributions

(a) Some common families of discrete distributions

Distribution	Possible values	Point probability	Restrictions	Mean	Variance	Probability generating function
Bernoulli	$r = 0$ or 1	$p^r(1-p)^{1-r}$	$0 < p < 1$	p	$p(1-p)$	$(1-p+ps)$
Binomial	$r = 0, 1, \ldots, n$	$\binom{n}{r}p^r(1-p)^{n-r}$	$0 < p < 1$	np	$np(1-p)$	$(1-p+ps)^n$
Poisson	$r = 0, 1, \ldots$	$e^{-\mu}\mu^r/r!$	$\mu > 0$	μ	μ	$e^{\mu(s-1)}$
Geometric	$r = 1, 2, \ldots$	$p(1-p)^{r-1}$	$0 < p < 1$	$1/p$	$(1-p)/p^2$	$ps/[1-(1-p)s]$
Geometric	$r = 0, 1, \ldots$	$p(1-p)^r$	$0 < p < 1$	$(1-p)/p$	$(1-p)/p^2$	$p/[1-(1-p)s]$
Negative binomial	$r = 0, 1, \ldots$	$\binom{k+r-1}{r}p^k(1-p)^r$	$0 < p < 1$	$k(1-p)/p$	$k(1-p)/p^2$	$p^k[1-(1-p)s]^{-k}$
Hypergeometric	$r = 0, 1, \ldots, \min(n, m_1)$	$\dfrac{\binom{m_1}{r}\binom{m-m_1}{n-r}}{\binom{m}{n}}$	—	$\dfrac{m_1 n}{m}$	$\dfrac{m_1 n(m-m_1)(m-n)}{m^2(m-1)}$	—

(b) Some common families of continuous distributions

Distribution	Possible values	Density function	Restrictions	Mean	Variance	General shape
Uniform	$a < x < b$	$1/(b-a)$	—	$(a+b)/2$	$(b-a)^2/12$	
Normal	$-\infty < x < \infty$	$\exp[-(x-\mu)^2/2\sigma^2]/\sqrt{(2\pi\sigma^2)}$	$\sigma > 0$	μ	σ^2	
Exponential	$x > 0$	$\lambda e^{-\lambda x}$	$\lambda > 0$	$1/\lambda$	$1/\lambda^2$	
Gamma	$x > 0$	$\lambda e^{-\lambda x}(\lambda x)^{r-1}/\Gamma(r)$	$\lambda > 0, r > 0$	r/λ	r/λ^2	for $r > 1$ (Gamma)
Weibull	$x > 0$	$m\lambda x^{m-1}\exp[-\lambda x^m]$	$m > 0, \lambda > 0$			for $m > 1$ (Weibull)

There are many other families of probability distributions which are applicable to particular problems. A catalogue of distributions (discrete, continuous and multivariate) and their properties is given, for example, by Johnson and Kotz (1969; 1971; 1972).

Multivariate distributions may be defined by the **joint probability function** in the discrete case (e.g. Prob($X = x$, $Y = y$) = probability that X takes value x and Y takes value y) or the **joint pdf** in the continuous case. The **multivariate normal** is a particularly useful family with some remarkable properties which can be illustrated for the bivariate case. Suppose the random variables (X, Y) are bivariate normal. This distribution is specified by five parameters, namely the mean and variance of each variable and the correlation coefficient of the two variables (section A.6). The formula for the bivariate normal is not particularly enlightening and will not be given here. It can be shown that the marginal distributions of X and Y are both univariate normal, as is the conditional distribution of X given $Y = y$ or of Y given $X = x$, for any value of y or of x.

If a vector random variable, $\mathbf{X}$, of length p, is multivariate normal, we write

$$\mathbf{X} \sim N_p(\boldsymbol{\mu}, \Sigma)$$

where $\boldsymbol{\mu}$ is the ($p \times 1$) mean vector of $\mathbf{X}$ and Σ is the ($p \times p$) covariance matrix of $\mathbf{X}$ whose (i, j)th element denotes the covariance between the ith and jth elements of $\mathbf{X}$, namely $E[(X_i - \mu_i)(X_j - \mu_j)]$ in an obvious notation. Here the expectation operator is extended to a function of two random variables in an 'obvious' way. For example, in the discrete case

$$E[g(X, Y)] = \sum_{x, y} g(x, y)P(X = x, Y = y).$$

Note that if X, Y are independent random variables, then Prob($X = x$, $Y = y$) = Prob($X = x$)Prob($Y = y$) and it can then be shown that $E[f(X)h(Y)] = E[f(X)]E[h(Y)]$. Another general useful rule for any two independent random variables is Var($X + Y$) = Var(X) + Var(Y).

An appropriate, entertaining end to this section is provided by W. J. Youden's illustration of the normal curve:

THE
NORMAL
LAW OF ERROR
STANDS OUT IN THE
EXPERIENCE OF MANKIND
AS ONE OF THE BROADEST
GENERALIZATIONS OF NATURAL
PHILOSOPHY ◆ IT SERVES AS THE
GUIDING INSTRUMENT IN RESEARCHES
IN THE PHYSICAL AND SOCIAL SCIENCES AND
IN MEDICINE AGRICULTURE AND ENGINEERING ◆
IT IS AN INDISPENSABLE TOOL FOR THE ANALYSIS AND THE
INTERPRETATION OF THE BASIC DATA OBTAINED BY OBSERVATION AND EXPERIMENT

– W. J. Youden

A.4 Estimation

The two main branches of **statistical inference** are **estimation** and **hypothesis testing**. The latter is considered in the next section, A.5.

Suppose we have a random sample $X_1, \ldots, X_n$ from a population whose distribution depends on an unknown parameter θ. A statistic is a function of the sample values, say $T(X_1, \ldots, X_n)$. One problem is to find a suitable statistic, say $\hat{\theta}(X_1, \ldots, X_n)$, which provides a good **point estimator** of θ. The realization of this statistic for a particular sample, say $\hat{\theta}(x_1, \ldots, x_n)$, is called a **point estimate**. A point estimator is **unbiased** if $E(\hat{\theta}) = \theta$. A statistic, T, is said to be **sufficient** if the conditional distribution of the sample given T does not depend on θ, so that the statistic contains all the information about θ in the sample. An estimator is said to be **consistent** if it tends to get closer and closer to the true value as the sample size increases and **efficient** if it has relatively low variance (these are not rigorous definitions!).

There are several general methods of finding point estimators. A **method of moments** estimator is found by equating the required number of population moments (which are functions of the unknown parameters) to the sample moments and solving the resulting equations. The **method of maximum likelihood** involves finding the joint probability (or joint pdf in the continuous case) of the data given the unknown parameter, θ, and then maximizing this likelihood function (or usually its logarithm for exponential families) with respect to θ. Note that the E–M algorithm is an iterative two-stage (the E-step stands for expectation and the M-step for maximization) computational procedure for deriving maximum-likelihood estimates when the observations are incomplete in some way. For example, there may be missing values (which must be missing at random) or some observations may be censored (e.g. Cox and Oakes, 1984, Chapter 11; Crowder *et al.*, 1991).

The **method of least squares** estimates the unknown parameters by minimizing the sum of squared deviations between the observed values and the fitted values obtained using the parameter values. The **method of least absolute deviations** (or L_1 estimation) minimizes the corresponding sum of absolute deviations. The latter approach can be analytically difficult but is increasingly 'easy' on a computer. Whichever approach is adopted (and there are several not mentioned here), the wide availability of computer programs means that less attention need be paid to the technical details.

In addition to point estimates, many packages also calculate **confidence intervals** which provide an interval within which the unknown parameter will lie with a prescribed confidence (or probability). Interval estimates are usually to be preferred to point estimates. Beware of interval estimates presented in the form $a \pm b$, where it is not clear if b is one standard error, or two standard errors, or gives a 95% confidence interval or what.

We consider one particular problem in detail, namely that of estimating the mean of a normal distribution. Suppose a random sample, size n, is taken from a normal distribution with mean μ and variance σ^2. Denote the sample mean and variance by $\bar{x}, s^2$, respectively. The sample mean is an intuitive point estimate of μ and we write

$$\hat{\mu} = \bar{x}$$

where the 'hat' $(\hat{})$ over μ means 'an estimate of'. This estimate arises from several different estimation approaches including maximum likelihood, least squares and the method of moments. It can be shown that the sampling distribution of the sample mean, which would be obtained by taking repeated samples of size n, is $N(\mu, \sigma^2/n)$, where $\sigma/\sqrt{n}$ is called the **standard error** of the sample mean. If σ is known (rare in practice), then it can be shown that the 95% confidence interval (CI) for μ is given by

$$\bar{x} \pm 1.96\sigma/\sqrt{n} \qquad (A.4.1)$$

while if σ is unknown, we use $s/\sqrt{n}$ as the estimated standard error of $\bar{x}$ to give

$$\bar{x} \pm t_{0.025, n-1}s/\sqrt{n} \qquad (A.4.2)$$

where t denotes the appropriate upper percentage point of a t-distribution with $n - 1$ degrees of freedom such that $2\frac{1}{2}\%$ of the distribution lies above it.

We consider two more problems in brief. The sample variance, s^2, is an unbiased point estimate of σ^2 and the 95% CI for σ^2 is given by the range

$$\frac{(n-1)s^2}{\chi^2_{0.025, n-1}} \quad \text{to} \quad \frac{(n-1)s^2}{\chi^2_{0.975, n-1}} \qquad (A.4.3)$$

The second situation involves the comparison of two groups of observations. A random sample, size n_1, is taken from $N(\mu_1, \sigma^2)$ and a second random sample, size n_2, is taken from $N(\mu_2, \sigma^2)$. Note that the population variances are assumed equal. If the sample means and variances are $\bar{x}_1, \bar{x}_2, s_1^2, s_2^2$ respectively, then $\bar{x}_1 - \bar{x}_2$ is a point estimate of $\mu_1 - \mu_2$, and its standard error is

$$\sqrt{\left[\frac{\sigma^2}{n_1} + \frac{\sigma^2}{n_2}\right]} = \sigma\sqrt{\left[\frac{1}{n_1} + \frac{1}{n_2}\right]}$$

Then a 95% CI for $\mu_1 - \mu_2$, assuming σ is known, is given by

$$(\bar{x}_1 - \bar{x}_2) \pm 1.96\sigma\sqrt{\left[\frac{1}{n_1} + \frac{1}{n_2}\right]} \qquad (A.4.4)$$

If σ is unknown, then the combined estimate of σ^2 is given by

$$s^2 = [(n_1 - 1)s_1^2 + (n_2 - 1)s_2^2]/(n_1 + n_2 - 2) \qquad (A.4.5)$$

and the 95% CI for $(\mu_1 - \mu_2)$ is

$$(\bar{x}_1 - \bar{x}_2) \pm t_{0.025, n_1 + n_2 - 2}s\sqrt{\left[\frac{1}{n_1} + \frac{1}{n_2}\right]} \qquad (A.4.6)$$

An important class of estimators contains those which are insensitive to departures from the assumptions on which the model is based. In particular, they usually accommodate (or even reject) outlying observations by giving them less weight than would otherwise be the case. A **trimmed mean**, for example, omits a specified percentage of the extreme observations while a **Winsorized mean** replaces outlying values with the nearest retained observation. One important class of robust estimators are **M-estimators** (see Hoaglin, Mosteller and Tukey, 1983, Chapters 9 to 12), which minimize an objective function which is more general than the familiar sum of squared residuals. For a location parameter θ, we minimize $\sum_{i=1}^{n} \rho(x_i - \theta)$ with respect to θ, where ρ is a suitably chosen function which is usually symmetric and differentiable. For example, $\rho(u) = u^2$ gives a least-squares estimate, but we arrange for $\rho(u)$ to increase at a lower rate for large u in order to achieve robustness. If $\psi(u) = d\rho(u)/du$, then the M-estimator is equivalently obtained by solving $\sum_{i=1}^{n} \psi(x_i - \theta) = 0$. Another class of estimators are **L-estimators** which are a weighted average of the sample order statistics (which are the sample values when arranged in order of magnitude). A trimmed mean is an example of an L-estimator.

There is growing interest in the use of estimation methods which involve **resampling** the observed data (e.g. Efron and Gong, 1983; Efron and Tibshirani, 1993). Given a particular estimator, resampling methods are intended to help the analyst assess its properties, particularly its bias and standard error, when the theoretical sampling distribution of the estimator is difficult to find or when parametric assumptions are difficult to justify. An old idea is to split the observed sample into a number of subgroups, to calculate the estimator for each subgroup and to use the variance of these quantities to estimate the variance of the overall estimator. The usual form of **jackknifing** is an extension of this idea. Given a sample of n observations, the observations are omitted one at a time, giving n (overlapping) groups of $n - 1$ observations. The estimator is calculated for each group, as well as for the whole sample, and these values can be used to assess the bias and standard error of the usual estimator as well as providing an estimator with first-order bias effects removed. An alternative way of reusing the sample is **bootstrapping**. The idea is to simulate the properties of a given estimator by taking repeated samples of size n with replacement from the observed empirical distribution in which $x_1, x_2, \ldots, x_n$ are each given probability mass $1/n$. (In contrast jackknifing generally takes n samples of size $n - 1$ **without** replacement.) Each bootstrap sample gives an estimate of the unknown population parameter. The average of these values is called the **bootstrap estimator**, and the variance is called the **bootstrap variance**. Sampling from the observed data is sometimes called **non-parametric bootstrapping**, whereas sampling from an assumed distribution by simulation is sometimes called **parametric bootstrapping**, though this term could cause confusion.

It is still not clear when, and how, bootstrapping can usefully be applied. While it is an intriguing, and potentially valuable, tool, it arguably has 'more potential for abuse than any other statistical technique yet invented' (Reese,

1994), and there is a real danger that it will be used to pretend that the sample size is larger than it actually is.

Bootstrapping and jackknifing are two examples of computationally intensive estimation methods. A third technique, called **cross-validation**, is a closer relative of jackknifing but is concerned with assessing the prediction error of a given model. Leaving out one (or more) observations at a time, a model is fitted to the remaining points and used to predict the deleted point(s). The resulting prediction errors can be used to provide an assessment of the model's predictive ability. The sum of squares of these errors can also be used to make a choice between different models as in the PRESS (predicted residual sum of squares) approach. The use of a **hold-out sample**, sometimes called **data splitting**, could be regarded as a special case of cross-validation, wherein part of the data is kept in reserve to test the model fitted to the remainder of the data. This device is much used in time-series analysis, but it is unclear how much data should be kept in reserve. A genuinely new (confirmatory) sample is much better for validating a model wherever this is possible.

The confidence intervals given earlier (expressions (A.4.1)–(A.4.6)) arise from the **frequentist** approach to statistical inference (section 7.4). **Bayesian inference** is a completely different approach to inference, with its own philosophy based on subjective probabilities. It depends on Bayes' theorem, which, as applied to inference, says that if H is some hypothesis or proposition (for example, that a model parameter takes a particular value) and D is some data, then Prob $(H|D)$ – the **posterior** probability of H – is proportional to Prob(H) Prob($D|H$), where the **prior** probability Prob(H) expresses our prior belief that H is true, and Prob $(D|H)$ is the **likelihood** of the data. This approach appeals to many statisticians, and there is a brief discussion of its advantages and difficulties in section 7.4 of Part One.

A.5 Significance tests

This section introduces the terminology of hypothesis testing and also describes some specific types of test. The difficulties and dangers of the procedures are discussed in section 7.2 of Part One.

A **hypothesis** is a conjecture about the population from which a given set of data are to be drawn. A significance test is a procedure for testing a particular hypothesis called the **null hypothesis**, which is customarily denoted by H_0. The test consists of deciding whether the data are consistent with H_0. The analyst should normally specify H_0 and an **alternative hypothesis**, denoted by H_1, before looking at the data. In particular this specifies whether a **one-tailed** test (where we are only interested in departures from H_0 'in one direction') or a **two-tailed** test is appropriate.

A suitable **test statistic** should be selected to show up departures from H_0. The sampling distribution of the test statistic, assuming that H_0 is true, should be known. The observed value of the test statistic should then be calculated from the data. The level of significance of the result (the P-value) is the probability of getting a test statistic which is as extreme, or more extreme, than the one observed, assuming that H_0 is true. If $P < 0.05$, we say that the result is significant at the 5% level and that we have some evidence to reject H_0. If $P < 0.01$, the result is significant at the 1% level and we have strong evidence to reject H_0. Some 'shading' is advisable in practice, given doubts about assumptions. Thus the values $P = 0.049$ and $P = 0.051$ should both be seen as on the borderline of significance rather than as 'significant' and 'not significant'. If $P > 0.05$, we accept H_0, or rather 'fail to reject' H_0 if there are still doubts about it (e.g. if P is only just greater than 0.05 or the sample size is small). Note that P is **not** the probability that H_0 is true.

An error of **type I** (or of the first kind) is said to occur when H_0 is rejected incorrectly (i.e. when H_0 is actually true) because of an extreme-looking sample. An error of **type II** occurs when we incorrectly fail to reject H_0 (when H_1 is actually true) because of a sample which happens to look consistent with H_0. The **power** of a test is the probability of correctly rejecting H_0 and so equals [1 − Prob (error of type II)]. With small sample sizes, the power may be disturbingly low, and this aspect of tests deserves more attention.

Some specific tests in brief are as follows:

A.5.1 Tests on a sample mean

Suppose we have a random sample size n from $N(\mu, \sigma^2)$ and wish to test

$$H_0 : \mu = k$$

against

$$\left. \begin{array}{l} H_1 : \mu > k \\ \text{or } H_1 : \mu < k \end{array} \right\} \text{one-tailed}$$

$$\text{or } H_1 : \mu \neq k \quad \text{two-tailed}$$

A suitable test statistic is

$$Z = (\bar{x} - k)\sqrt{n}/\sigma \quad \text{if } \sigma \text{ is known} \tag{A.5.1}$$

or

$$t = (\bar{x} - k)\sqrt{n}/s \quad \text{if } \sigma \text{ is unknown} \tag{A.5.2}$$

where $\bar{x}, s$ are the sample mean and standard deviation. If H_0 is true, then $Z \sim N(0, 1)$ and $t \sim t_{n-1}$, the latter statistic giving rise to what is called a t-test. Note the assumptions that observations are independent and (approximately) normally distributed. If these assumptions are unreasonable, then it may be safer to carry out a non-parametric test (section A.5.5), though most parametric tests are robust to moderate departures from distributional assumptions.

A.5.2. Comparing two groups of observations

First decide if a two-sample test or a paired comparison test is appropriate. The latter test arises when each observation in one group has a natural paired observation in the other group so that a one-sample test can be carried out on the paired differences. Then $\bar{x}$ and s in formula (A.5.2) are the mean and standard deviation of the differences. The null hypothesis is usually that the population mean difference is zero (so that k is zero in (A.5.2)).

A two-sample t-test is illustrated in Exercise B.1.

A.5.3 Analysis of variance (ANOVA)

The one-way ANOVA generalizes the two-sample t-test to compare more than two groups and is described later in section A.7. More generally, ANOVA can be used to test the effects of different influences in more complicated data structures.

ANOVA generally involves one or more F-tests to compare estimates of variance. If s_1^2, s_2^2 are two independent estimates of variance based on v_1, v_2 degrees of freedom (DF), respectively, then the ratio s_1^2/s_2^2 may be compared with the appropriate percentage points of an F-distribution with (v_1, v_2) DF to test the hypothesis that the underlying population variances are equal.

A.5.4 The chi-squared goodness-of-fit test

This is applicable to frequency or count data. The observed frequencies in different categories are compared with the expected frequencies which are calculated assuming a given null hypothesis to be true. The general form of the test statistic is

$$\chi^2 = \sum_{\text{all categories}} \frac{(\text{observed frequency} - \text{expected frequency})^2}{\text{expected frequency}} \quad \text{(A.5.3)}$$

If H_0 is true, then this test statistic has an approximate χ^2-distribution whose degrees of freedom are (number of categories $- 1 -$ number of independent parameters estimated from the data). The test is illustrated in Exercises B.4, E.4, G.2, G.3, G.7(b). Note that the χ^2-test is always one-tailed as all deviations from a given H_0 will lead to 'large' values of χ^2. Also note that categories with 'small' expected frequencies (e.g. less than 5) may need to be combined in a suitable way. When a significant result is obtained, the analyst should inspect the differences between observed and expected frequencies to see *how* H_0 is untrue.

The analysis of categorical data is a wide subject of which the χ^2-test is just a part. There are, for example, special types of correlation coefficient for measuring the strength of dependence between two categorical variables with frequency data recorded in a two-way contingency table. A log-linear model (see section A.9) may be used to model systematic effects in structured frequency data.

A.5.5 Non-parametric (or distribution-free) tests

Non-parametric tests make as few assumptions as possible about the underlying distribution. The simplest test of this type is the **sign** test which essentially looks only at the sign (positive, negative, or zero for a 'tie') of differences. This throws away some information and is therefore not efficient, but it can be useful in an IDA as it can often be done 'by eye'. For example, in a paired comparison test the analyst can look at the signs of the differences. If they nearly all have the same sign, then there is strong evidence to reject the null hypothesis that the true average difference is zero. The binomial distribution can be used to assess results by assuming that the probability of a positive difference under H_0 is 1/2, ignoring ties.

It is more efficient to look at the magnitude of differences as well as the sign. Many tests are based on **ranks**, whereby the smallest observation in a sample is given rank 1, and so on. Equal observations are given the appropriate average rank. The **Wilcoxon signed-rank test** is a substitute for the one-sample t-test and is particularly suitable for paired differences. The absolute values of the differences, ignoring the signs, are ranked in order of magnitude. Then the signs are restored to the rankings and the sums of the positive rankings and of the negative rankings are found. The smaller of those two sums is usually taken as the test statistic and may be referred to an appropriate table of critical values. Values of the test statistics less than or equal to the critical value imply rejection of the null hypothesis that the median difference is zero.

The two-sample **Wilcoxon rank-sum test** (or equivalently the **Mann–Whitney U-test**) is a substitute for the two-sample t-test for testing that two populations have the same median. The two samples are combined to give a single group and then ranks are assigned to all the observations. The two samples are then reseparated and the sum of the ranks for each sample is found. The smaller of these two sums is usually taken as the test statistic and may be referred to tables of critical values. The equivalent Mann–Whitney approach orders all the observations in a single group and counts the number of observations in sample A that precede each observation in sample B. The U-statistic is the sum of these counts and may also be referred to a table of critical values. The **Kruskal–Wallis test** is the generalization of this test for comparing $k(>2)$ samples and is therefore the non-parametric equivalent of a one-way ANOVA.

There are various other non-parametric tests, and the reader is referred to a specialized book such as Sprent (1993).

When should a non-parametric approach be used? They are widely used in the social sciences, particularly in psychology, where data are often skewed or otherwise non-normal. They can be rather tricky to perform by hand (try finding a rank sum manually!) but are easy enough to perform using a computer package. They also have nice theoretical properties because they are often nearly as efficient as the corresponding parametric approach even when the parametric assumptions are true and they can be far more efficient when they are not. Despite this, they are often avoided in some scientific areas, perhaps because test statistics based on means and variances are intuitively more meaningful than quantities like rank sums.

A.6 Regression

Regression techniques seek to establish a relationship between a response variable, y, and one or more explanatory (or predictor) variables $x_1, x_2, \ldots$. The approach is widely used and can be useful, but is also widely misused. As in all statistics, if you fit a silly model, you will get silly results. Guard against this by plotting the data, using background information and past empirical evidence. I have had some nasty experiences with data where the explanatory variables were correlated, particularly with time-series data where successive observations on the same variable may also be correlated. Nowadays, computers make it easy to fit regression models and attention needs to focus on such questions as which variables to include, what to do about outliers, and so on (Exercise C.4).

A.6.1 Preliminary questions

Here are some questions which should be asked at the outset:

- Why do you want to fit a regression model anyway? What are you going to do with it when you get it? Have models been fitted beforehand to other similar data sets?
- How were the data collected? Are the x-values controlled by the experimenter and do they cover a reasonable range?

Begin the analysis, as usual, with an IDA, to explore the main features of the data. In particular, plot scatter diagrams to get a rough idea of the relationship, if any, between y and each x.

- Are there obvious outliers? Is the relationship linear? Is there guidance on secondary assumptions such as normality and whether the conditional 'error' variance is constant?

This is straightforward with just one x-variable, but with two or more x-variables you should be aware of the potential dangers of collapsing multivariate data onto two dimensions when the x-variables are correlated.

A.6.2 The linear regression model

The simplest case arises with one predictor variable, say x, where the scatter diagram indicates a linear relationship. The conditional distribution of the response variable, y, for a given fixed value of x has a mean value denoted by $E(y|x)$. The **regression curve** is the line joining these conditional expectations and is here assumed to be of the form:

$$E(y|x) = \alpha + \beta x \qquad (A.6.1)$$

where α is called the **intercept** and β is the **slope**. The deviations from this line are usually assumed to be independent, normally distributed with zero mean and constant variance, σ^2. These are a lot of assumptions!

Given n pairs of observations, namely $(x_1, y_1), \ldots, (x_n, y_n)$, the least-squares estimates of α and β are obtained by minimizing

$$\Sigma(\text{observed value of } y - \text{fitted value})^2 = \sum_{i=1}^{n} (y_i - \alpha - \beta x_i)^2$$

This gives

$$\hat{\alpha} = \bar{y} - \hat{\beta}\bar{x}$$

$$\hat{\beta} = \Sigma(x_i - \bar{x})(y_i - \bar{y})/\Sigma(x_i - \bar{x})^2$$

Note that $\hat{\beta}$ may be expressed in several equivalent forms such as $\Sigma(x_i - \bar{x})y_i/\Sigma(x_i - \bar{x})^2$, since $\Sigma(x_i - \bar{x}) = \bar{y}\Sigma(x_i - \bar{x}) = 0$.

Having fitted a straight line, the residual sum of squares is given by $\Sigma(y_i - \hat{\alpha} - \hat{\beta}x_i)^2$, which can be shown to equal $(S_{yy} - \hat{\beta}^2 S_{xx})$, where S_{yy}, for example, is the total corrected sum of squares of the y's, namely $\Sigma(y_i - \bar{y})^2$. The residual variance, σ^2, is usually estimated for all regression models by $s^2 = $ residual SS/residual DF = residual mean square which can readily be found in the output from most regression packages, usually in the ANOVA table (Table A.6.1). The residual DF is given generally by ($n-$ number of estimated parameters), which for linear regression is equal to $n - 2$.

Other useful formulae include:

(i) $100(1 - \alpha_0)\%$ CI for α is $\hat{\alpha} \pm t_{\alpha_0/2, n-2}\, s \sqrt{\left[\dfrac{1}{n} + \dfrac{\bar{x}^2}{\Sigma(x_i - \bar{x})^2}\right]}$.

(*Note:* The probability associated with a confidence interval (CI) is usually denoted by α, but we use α_0 where necessary to avoid confusion with the intercept, α.)

(ii) $100(1 - \alpha)\%$ CI for β is $\hat{\beta} \pm t_{\alpha/2, n-2} s / \sqrt{\Sigma(x_i - \bar{x})^2}$.

(iii) $100(1 - \alpha_0)\%$ CI for $\alpha + \beta x_0$ is $\hat{\alpha} + \hat{\beta} x_0 \pm t_{\alpha_0/2, n-2} s \sqrt{\left[\frac{1}{n} + \frac{(x_0 - \bar{x})^2}{\Sigma(x_i - \bar{x})^2} \right]}$.

The latter CI is for the mean value of y, given $x = x_0$. Students often confuse this with the **prediction interval** for y, for which there is a probability $(1 - \alpha_0)$ that a future single observation on y (not the mean value!), given $x = x_0$, will lie in the interval

$$\hat{\alpha} + \hat{\beta} x_0 \pm t_{\alpha_0/2, n-2} s \sqrt{\left[1 + \frac{1}{n} + \frac{(x_0 - \bar{x})^2}{\Sigma(x_i - \bar{x})^2} \right]}.$$

The analysis of variance (ANOVA) partitions the total variability in the y-values into the portion explained by the linear model and the residual, unexplained variation. The ANOVA table is shown in Table A.6.1.

Table A.6.1 ANOVA table for linear regression

Source	SS	DF	MS	E(MS)
Regression	$\hat{\beta}^2 S_{xx}$	1	$\hat{\beta}^2 S_{xx}$	$\sigma^2 + \beta^2 S_{xx}$
Residual	by subtraction	$n - 2$	s^2	σ^2
Total	S_{yy}	$n - 1$		

Numerous tests of significance are possible and many are performed routinely in computer output. Note that many arise in different, but equivalent, forms. For example, to test if the true slope of the line is zero (i.e. is it worth fitting a line at all?), you can perform a t-test on $\hat{\beta}$, an F-test on the ratio of the regression MS to the residual MS, or simply see if the above CI for β includes the value zero.

Mean-corrected model

There are some computational advantages in using a mean-corrected form of (A.6.1), namely

$$E(y|x) = \alpha^* + \beta(x - \bar{x}). \tag{A.6.2}$$

The slope parameter is unchanged, while α and α^* are related by $\alpha^* - \beta\bar{x} = \alpha$. Then $\hat{\alpha}^* = \bar{y}$ and the fitted line is of the form $y - \bar{y} = \hat{\beta}(x - \bar{x})$. Mean-corrected models are used routinely in many forms of regression.

A.6.3 Curvilinear models

A linear model may be inappropriate for external reasons or because the eye detects non-linearity in the scatter plot. One possibility is to transform one or both variables so that the transformed relationship is linear. The alternative is to fit a non-linear curve directly. The commonest class of curvilinear models are polynomials such as the quadratic regression curve

$$E(y|x) = \alpha + \beta_1 x + \beta_2 x^2.$$

Similar assumptions about 'errors' are usually made as in the linear case. Polynomial models can be fitted readily by most computer packages.

A.6.4 Non-linear models

These are usually defined to be **non-linear in the parameters** and are thus distinct from curvilinear models. An example is

$$E(y|x) = 1/(1 + \theta x).$$

Models of this type are trickier to handle (e.g. Draper and Smith, 1981, Chapter 10; Ratkowsky, 1983; Bates and Watts, 1988). Whereas linear models can be fitted by least squares using matrix manipulation, non-linear models are much harder to fit even when it is reasonable to assume that errors are independently and identically distributed $N(0, \sigma^2)$ so that least squares can be used. It will usually be necessary to solve a set of simultaneous non-linear equations for which some sort of numerical optimization will be needed. Convergence problems sometimes arise and it can sometimes help to reparametrize the model. Non-linear models arise widely, particularly in agricultural and chemical applications, and can readily be fitted nowadays using general packages such as GENSTAT or SAS or a specialist package such as MLP (Ross, 1990).

A.6.5 Multiple regression

With k explanatory variables, the multiple linear regression model is

$$E(y|x_1, \ldots, x_k) = \alpha + \beta_1 x_1 + \cdots + \beta_k x_k$$

together with the usual assumptions about the errors. If the data are given by $(y_1, x_{11}, \ldots, x_{k1}), \ldots, (y_n, x_{1n}, \ldots, x_{kn})$, the model may easily be fitted by least squares using a computer. Curvilinear terms may be introduced, for example, by letting $x_2 = x_1^2$. The x-variables are usually centred (or mean-corrected), and numerical considerations also suggest scaling the x-variables to have equal variance, by considering $(x_i - \bar{x}_i)/s_i$ where s_i is the standard deviation of observed values of x_i. Then the fitted slope is scaled in an obvious way but other expressions, such as sums of squares of the y-values, are unchanged. More complicated transformations of the explanatory variables, such as the Box–Cox transformation (see section 6.8 of Part One), will occasionally be needed.

Estimates of $\beta_1, \ldots, \beta_k$ will only be uncorrelated if the design is **orthogonal**. This will happen if the x-values are chosen to lie on a symmetric, regular grid, or more mathematically if

$$\sum_{j=1}^{n} (x_{sj} - \bar{x}_s)(x_{tj} - \bar{x}_t) = 0 \qquad \text{for all } s, t \text{ such that } s \neq t.$$

When the x-variables can be controlled, it is helpful to choose them so as to get an orthogonal design. This not only simplifies the analysis but also enables the effect of each x-variable to be assessed independently of the others. It is also desirable to randomize the order of the experiments so as to eliminate the effects of nuisance factors.

However, in practice multiple regression is more often used on observational data where the x-variables are correlated with each other. In the past, explanatory variables were often called independent variables, but this misleading description has now been largely abandoned. With correlated x-variables, it is not safe to try and interpret individual coefficients in the fitted model, and the fitted model may be misleading despite appearing to give a good fit (Exercise C.3). Sometimes the x-variables are so highly correlated that the data matrix gives rise to a matrix of sums of squares and cross-products (section A.8) which is ill-conditioned (or nearly singular). Then it may be wise to omit one or more suitably chosen x's, or consider the use of special numerical procedures, such as ridge regression. Multicollinearity problems (e.g. Wetherill, 1986, Chapter 4) arise because of near or exact linear dependencies amongst the explanatory variables. They may be caused by the inclusion of redundant variables, by physical constraints or by the sampling techniques employed.

With time-series data, particularly those arising in economics, there may be correlations not only between different series, but also between successive values of the same series (called **autocorrelation**). This provides a further complication. A multiple regression model may then include lagged values of the response variable (called **autoregressive** terms) as well as lagged values of the explanatory variables.

Some explanatory variables may contribute little or nothing to the fit and need to be discarded. Choosing a subset of the x-variables may be achieved by a variety of methods including **backward elimination** (where the least important variable is successively removed until all the remaining variables are significant) or **forward selection** (where the procedure begins with no x-variables included). Sometimes there are several alternative models, involving different x-variables, which fit the data almost equally well. Then it is better to choose between them using external knowledge where possible, rather than relying completely on automatic variable selection. In particular, there may be prior information about the model structure which suggests that some variables must be included. It is often tempting to begin by including a large number of x-variables. Although the fit may appear to improve, it may be spurious in that the fitted model has poor predictive performance over a range of conditions. As a crude rule of thumb, I generally suggest that the number of variables should not exceed one quarter of the number of observations and should preferably not exceed about four or five. In an exploratory study it is perhaps reasonable to include rather more variables just to see which are important – but do not believe the resulting fitted equation without checking on other data sets.

It has been suggested (Preece, 1984) that there are about 100 000 multiple regressions carried out each day, of which only 1 in 100 are sensible. While this may be a slight exaggeration, it does indicate the overuse of the technique and statisticians should be just as concerned with the silly applications as with the sensible ones.

A.6.6 Coefficient of determination

This is useful for assessing the fit of all types of regression model and is usually defined by

$$R^2 = \text{explained SS/total SS}.$$

The total (corrected) SS, namely $\Sigma(y_i - \bar{y})^2$, is partitioned by an ANOVA into the explained (or regression) SS and the residual SS where residual SS = $\Sigma(\text{observed } y - \text{fitted } y)^2$. Thus R^2 must lie between 0 and 1. The better the fit, the closer will R^2 lie towards one.

In simple linear regression, it can be shown that $R^2 = (\text{correlation coefficient})^2$ (see below). More generally, R^2 is the square of the correlation between the observed and fitted values of y. Thus R is sometimes called the **multiple correlation coefficient**.

One problem with interpreting R^2 is that it always gets larger as more variables are added, even if the latter are of no real value. An alternative coefficient produced by many packages is R^2 (adjusted) which adjusts the value of R^2 to take the number of fitted parameters into account. Instead of

$$R^2 = 1 - \frac{\text{residual SS}}{\text{total SS}}$$

calculate

$$R^2(\text{adjusted}) = 1 - \frac{\text{residual MS}}{\text{total MS}}$$

which is always smaller than R^2 and less likely to be misleading.

A 'high' value of R^2 is commonly taken to mean 'a good fit' and many people are impressed by values exceeding 0.8. Unfortunately, it is very easy to get values exceeding 0.99 for time-series data which are quite spurious. Armstrong (1985, p. 487) gives a delightful set of rules for 'cheating', so as to obtain a high value of R^2. They include the omission of outliers, the inclusion of lots of variables and the use of R^2 rather than R^2 (adjusted).

A.6.7 Model checking

After fitting a regression model, it is important to carry out appropriate diagnostic checks on the residuals (cf. section 5.3.3 of Part One). If outliers are present, then they may be adjusted or removed, or alternatively some form of robust regression could be used. More generally, it is of interest to detect **influential** observations whose deletion results in substantial changes to the fitted model. Formulae regarding residuals, influence, outliers and leverage can be presented more conveniently in the matrix notation of the general linear model and will therefore be deferred to section A.8.

It may also be necessary to test the data for normality (Wetherill, 1986, Chapter 8) and for constant variance (Wetherill, 1986, Chapter 9). If the conditional variance is found not to be constant, then the data are said to be **heteroscedastic** and it may be appropriate to fit a model by weighted least squares where the less accurate observations are given less weight (section A.8). For example, in linear regression it may be reasonable to assume that the conditional variance increases linearly with the value of the explanatory variable.

A.6.8 Correlation coefficient

This is a (dimensionless) measure of the linear association between two variables. The usual **product moment** correlation coefficient is given by

$$r = \frac{\Sigma(x_i - \bar{x})(y_i - \bar{y})}{\sqrt{[\Sigma(x_i - \bar{x})^2 \Sigma(y_i - \bar{y})^2]}}$$

It can be shown that $-1 \le r \le +1$. If a linear regression model is fitted then it can be shown that

$$r^2 = 1 - \frac{\text{residual SS}}{\text{total SS}} = \text{coefficient of determination}$$

Other measures of correlation are available for other types of data, such as discrete and ranked data. An example is **Spearman's rank correlation coefficient**, which is given by

$$r = 1 - 6\Sigma d_i^2 / n(n^2 - 1)$$

where d_i is the difference in the rankings of the ith x- and y-observations. Rank correlations require less in the way of assumptions than product-moment correlation and should probably be used more often than they are.

Tables of critical values are available to help decide which correlations are significantly large, but an adequate approximation for most purposes is that value outsides the range $\pm 2/\sqrt{n}$ are significant. It is harder to say how large a correlation needs to be in order to be judged 'interesting'. For example, if $|r| < 0.3$, then the fitted line explains less than 10% of the variation ($r^2 = 0.09$) and will probably be of little interest even if the sample size is large enough to make it significant. More generally it can be hard to assess and interpret correlations, as illustrated by Exercises C.1 and C.2. Finally, we note that the most common mistake in interpreting 'large' correlations is to suppose that they demonstrate a cause-and-effect relationship.

Although a correlation coefficient can be calculated when x is a controlled variable, it is usually more meaningful when both x and y are random variables whose joint distribution is bivariate normal. Then r provides a sensible estimate of the **population correlation coefficient** which is usually denoted by ρ. In the bivariate normal case it can also be shown that the regression curves of y on x and of x on y are both straight lines and can be estimated by the regression techniques described above. To estimate the regression curve of x on y, all formulae are 'reversed', changing x to y and vice versa. The larger the correlation, the smaller will be the angle between the two regression lines.

A.6.9 Logistic regression

This is a special type of regression which may be appropriate when the response variable is binary. It is briefly discussed in section A.9.

A.6.10 Non-parametric regression

There is much current interest in this form of regression where the observations are assumed to satisfy

$$y_i = g(x_i) + \varepsilon_i$$

and the form of the function g is determined from the data by smoothing rather than by being specified beforehand. The residual variation is usually assumed to have constant variance (although this assumption can be relaxed). There are several approaches which generally trade off goodness-of-fit with some measure of smoothness. For example, the **spline smoothing** approach (e.g. Green and Silverman, 1993) chooses g so as to minimize $\Sigma[y_i - g(x_i)]^2$ in conjunction with a 'roughness' penalty, depending on the second derivative of g, which ensures that g is 'reasonably smooth'. This leads to a function, g called a **cubic spline** which has the properties that it is a cubic polynomial in each interval (x_i, x_{i+1})

and that g and its first two derivatives are continuous at each x_i. The x_i-values are called **knots**.

A.6.11 Calibration

Regression is concerned with predicting y for a given value of x – often called the **prediction problem**. The reverse problem, called the **calibration problem**, is to decide which value of x leads to a specified mean value of y, say y_0. The 'obvious' classical estimator in the linear case is given by $x = (y_0 - \hat{\alpha})/\hat{\beta}$, but note that this is not an unbiased estimator. There are several alternative approaches (e.g. Miller, 1986).

Further reading

Most textbooks provide an introduction to regression. Weisberg (1985) provides a more thorough introduction. Wetherill (1986) deals with many of the practical problems involved in multiple regression. Draper and Smith (1981) is the acknowledged reference text on regression. Cook and Weisberg (1982) and Atkinson (1985) provide a detailed treatment of residuals and influence in regression.

The following poem provides a salutary end to this section.

The Ballade of Multiple Regression

If you want to deal best with your questions,
Use multi-regression techniques;
A computer can do in a minute
What, otherwise done, would take weeks.
For 'predictor selection' procedures
Will pick just the ones best for you
And provide the best-fitting equation
For the data you've fitted it to.

But did you collect the right data?
Were there 'glaring omissions' in yours?
Have the ones that score highly much meaning?
Can you tell the effect from the cause?
Are your 'cause' factors ones you can act on?
If not, you've got more work to do;
Your equation's as good – or as bad – as
The data you've fitted it to.

Tom Corlett, *Applied Statistics*,
1963, **12**, p. 145 (first two verses only)

A.7. Analysis of variance (ANOVA)

ANOVA is a general technique for partitioning the overall variability in a set of observations into components due to specified influences and to random error (or haphazard variation). The resulting ANOVA table provides a concise summary of the structure of the data and a descriptive picture of the different sources of variation. In particular, an estimate of the error variance (the residual mean square) is produced. For a general linear model with normal errors (section A.8) this in turn allows the computation of CIs when estimating the effects of the explanatory influences, and also the testing of hypotheses about them, usually by means of F-tests. ANOVA can be applied to experimental designs of varying complexity, and we have already seen it applied to a linear regression model (section A.6) where the explanatory influence is the effect of the predictor variable.

Here we consider the 'one-way' case in detail and refer briefly to more complicated designs. Suppose we have k groups of observations with n_i observations in group i such that $n = \sum n_i$, and let y_{ij} denote the jth observation in group i. Denote the ith group mean by $\bar{y}_i = \sum_j y_{ij}/n_i$ and the overall mean by $\bar{y}$. The purpose of the experiment is probably to assess the differences between groups. (How large are the differences? Are they significantly large?) In order to do this, we start with an IDA, calculating summary statistics and plotting a series of boxplots to compare groups. If the results are not obvious, or if an estimate of the residual variation is required to calculate CIs for effects of interest, then a formal ANOVA is needed which compares the variability between groups with the variability within groups. The total corrected sum of squares of the y-values, namely $\sum_{i,j}(y_{ij} - \bar{y})^2$, is partitioned into the between-groups SS and the residual within-groups SS as in the ANOVA Table A.7.1. By writing the 'total' deviation

$$y_{ij} - \bar{y} = (y_{ij} - \bar{y}_i) + (\bar{y}_i - \bar{y})$$

as the sum of a 'within' and a 'between' deviation, squaring and summing over all i, j, it can be shown that the sums of squares 'add up'. Table A.7.1

Table A.7.1 One-way ANOVA table

Source of variation	SS	DF	MS
Between-groups	$\sum_i n_i(\bar{y}_i - \bar{y})^2$	$k - 1$	s_B^2
Within-groups (or residual)	$\sum_{i,j}(y_{ij} - \bar{y}_i)^2$	$n - k$	s^2
Total	$\sum_{i,j}(y_{ij} - \bar{y})^2$	$n - 1$	

also shows the appropriate degrees of freedom (DF) and the mean squares (MS = SS/DF). The residual MS, s^2, provides an estimate of the underlying residual variance and can be used in an F-test (to see if the between-group MS, s_B^2, is significantly large) and/or to estimate confidence intervals for differences between group means. (This emphasizes that ANOVA is not just used for testing hypotheses.)

The F-test is based on the ratio s_B^2/s^2 and assumes the following model:

$$y_{ij} = \mu + t_i + \varepsilon_{ij} \qquad (i = 1, \ldots, k; j = 1, \ldots, n_i) \qquad \text{(A.7.1)}$$

where μ is the overall mean, t_i the effect of the ith group, and ε_{ij} the random error for the jth observation in group i.

The errors are assumed to be normally distributed with mean zero and constant variance, σ^2, and also to be independent. Note the large number of assumptions! If the group effects $\{t_i\}$ are regarded as fixed, with $\Sigma t_i = 0$, then we have what is called a **fixed-effects** model. However, if the $\{t_i\}$ are assumed to be a random sample from $N(0, \sigma_t^2)$, then we have what is called a **random-effects** model (or **variance-components** model).

In a one-way ANOVA, the null hypothesis for the fixed-effects model is that there is no difference between groups so that all the t_i are zero. Then $E(s_B^2) = E(s^2) = \sigma^2$ under H_0. H_0 is rejected at the 5% level if the observed F-ratio $= s_B^2/s^2$ is significantly large compared with $F_{0.05,k-1,n-k}$.

A point estimate for t_i is given by $\bar{y}_i - \bar{y}$. Perhaps of more importance are the differences between groups and a point estimate of, say, $t_1 - t_2$ is given by $\bar{y}_1 - \bar{y}_2$. Confidence intervals may also be found using $s/\sqrt{n_i}$ as the estimated standard error of $\bar{y}_i$. The general process of seeing which group means differ significantly from each other is referred to as making **multiple comparisons**. The simplest method, called the **least significant difference** approach, should only be used when the ANOVA F-test is significant. It says that the means of groups i and j differ significantly (at the 5% level), if their absolute difference exceeds

$$t_{0.025,n-k}s\sqrt{\left(\frac{1}{n_i} + \frac{1}{n_j}\right)}$$

A two-way ANOVA is appropriate for data such as those arising from a randomized block design (section A.11) where there is one observation on each of k treatments in each of r blocks and model (A.7.1) is extended to

$$y_{ij} = \mu + t_i + b_j + \varepsilon_{ij} \ (i = 1, \ldots, k; j = 1, \ldots, r) \qquad \text{(A.7.2)}$$

where b_j denotes the effect of the jth block. Then the total variation is partitioned into sums of squares due to treatments, to blocks and to the residual variation.

In model (A.7.2) the t_i and b_j are sometimes referred to as the **main effects**. In a replicated complete factorial experiment (section A.11), it is also possible to estimate **interaction** terms which measure the joint effect of the levels of two or more effects. Then model (A.7.2) would be extended to include terms of the form γ_{ij}, which denotes the joint effect of factor I at the ith level and factor II at the jth level. If there are no interaction terms, the two factors are said to be **additive**.

It is impossible and unnecessary to give detailed formulae for all the many types of design which can arise. What is important is that you should:

1. understand the main types of experimental design (section A.11).
2. understand the underlying model which is applicable to the given data structure – for example, you should know if a fixed-effects or random-effects model is appropriate and know which main effects and/or interaction terms are included. You should also understand the assumptions being made about the residual variation.
3. be able to interpret computer output; in particular you should be able to pick out the residual mean square and understand its key role in estimation and significance testing.

The use of ANOVA is illustrated in several exercises in Part Two including B.2 (a one-way ANOVA), B.9 (a one-way ANOVA after transformation), E.1 (a two-way ANOVA), E.2 (a three-way ANOVA for a Latin square) and E.3 (an unbalanced two-way ANOVA).

Further reading

Numerous textbooks cover ANOVA. One rather unusual book (described as 'the confessions of a practising statistician'), which gives much useful advice is that by Miller (1986).

A.8 The general linear model

This general class of models includes regression and ANOVA models as special cases. By using matrix notation, many general results may be expressed in a relatively simple way.

Let $\mathbf{y}$ denote the $(n \times 1)$ vector of observations on a response variable y, and $\boldsymbol{\beta}$ denote a $(p \times 1)$ vector of (usually unknown) parameters. Then the general linear model can be written as

$$\mathbf{y} = X\boldsymbol{\beta} + \mathbf{e} \tag{A.8.1}$$

where X is an $(n \times p)$ matrix of known quantities and $\mathbf{e}$ is an $(n \times 1)$ vector of random error terms. Note that each element of $\mathbf{y}$ is a linear combination of the (unknown) parameters plus an additive error term.

In regression, the elements of X will include the observed values of the k predictor variables. If the regression model also includes a constant term, then X will include a column of ones and we have $p = k + 1$. In ANOVA the elements of X are chosen to include or exclude the appropriate parameters for each observation and so are usually 0 or 1. Each column of X can then be regarded as an **indicator variable** and X is usually called the **design matrix**. In the **analysis of covariance**, which can be regarded as a mixture of regression and ANOVA, X will include a mixture of predictor and indicator variables. Here the predictor variables are sometimes called **covariates** or **concomitant variables**.

It is often assumed that the elements of **e** are independent normally distributed with zero mean and constant variance, σ^2. Equivalently, using the multivariate normal distribution, we can write $\mathbf{e} \sim N(\mathbf{0}, \sigma^2 I_n)$ where $\mathbf{0}$ is an $(n \times 1)$ vector of zeros, I_n is the $(n \times n)$ identity matrix and $\sigma^2 I_n$ is the variance-covariance matrix of **e**.

The least squares estimate $\hat{\boldsymbol{\beta}}$ of $\boldsymbol{\beta}$ is chosen to minimize the residual sum of squares and is obtained by solving the normal equations

$$(X^T X)\hat{\boldsymbol{\beta}} = X^T \mathbf{y} \tag{A.8.2}$$

If X is of full rank p (assuming $n > p$), then $(X^T X)$ is square, symmetric and non-singular and can be inverted to give

$$\hat{\boldsymbol{\beta}} = (X^T X)^{-1} X^T \mathbf{y} \tag{A.8.3}$$

The Gauss–Markov theorem says that these estimates are the best (i.e. minimum-variance) linear unbiased estimates. The least-squares estimates are also maximum-likelihood estimates if the errors are normally distributed.

The generalized least-squares estimate of $\boldsymbol{\beta}$ is appropriate when the variance-covariance matrix of **e** is of the more general form $\sigma^2 \Sigma$, where Σ is $(n \times n)$ symmetric and positive definite. Then (A.8.3) becomes

$$\hat{\boldsymbol{\beta}} = (X^T \Sigma^{-1} X)^{-1} X^T \Sigma^{-1} \mathbf{y}. \tag{A.8.4}$$

In particular, if Σ is diagonal so that the errors are uncorrelated but perhaps have unequal variances, then we have **weighted least squares**. Note that (A.8.3) is a special case of (A.8.4) and that $\text{Var}(\hat{\boldsymbol{\beta}}) = \sigma^2 (X^T \Sigma^{-1} X)^{-1}$.

In regression, the solution of (A.8.3) and (A.8.4) is numerically more stable if mean-corrected values of the predictor variables are used (equation A.6.2), and then $X^T X$ is called the **corrected cross-product matrix**. It may also help to scale the predictor variables to have unit variance. The solution of (A.8.3) and (A.8.4) will become unstable if the predictor variables are highly correlated, so that there is near linear dependence between them. Then $X^T X$ may be ill-conditioned (or nearly singular) so that there is difficulty in finding its inverse. Ways of overcoming this problem were discussed briefly in section A.6.

We return now to the case where $\text{Var}(\mathbf{e}) = \sigma^2 I_n$. Then the vector of fitted values, $\hat{\mathbf{y}}$ is given by

$$\hat{\mathbf{y}} = X\hat{\boldsymbol{\beta}}$$
$$= X(X^T X)^{-1} X^T \mathbf{y}$$
$$= H\mathbf{y} \qquad\qquad\qquad (A.8.5)$$

where $H = X(X^T X)^{-1} X^T$ is called the **hat matrix** because it predicts the fitted (hat) values of y from the observed values. The vector of raw residuals is given by

$$\hat{\mathbf{e}} = (I_n - H)\mathbf{y}$$

and an unbiased estimate of σ^2 is given by

$$\hat{\sigma}^2 = \hat{\mathbf{e}}^T \hat{\mathbf{e}}/(n - p)$$
$$= (\text{residual SS})/(\text{residual DF})$$

The diagonal elements of H, namely $\{h_{ij}\}$, are useful in a variety of ways. The effect of the ith observation is more likely to be 'large' if h_{ii} is 'large' and so h_{ii} is called the **leverage** or **potential** of the ith observation. The raw residuals can be misleading because they have different standard errors. Most computer packages therefore standardize to give **scaled** or **studentized residuals**, which will have common variance equal to one if the model is correct. The **internally** studentized residuals are given by

$$r_i = \hat{e}_i/[\hat{\sigma}\sqrt{(1 - h_{ii})}]$$

for $i = 2, \ldots, n$ where $\hat{e}_i$ is the raw residual for the ith obervation and $\hat{\sigma}$ is the overall estimate of σ. The **externally** studentized residuals are given by

$$t_i = \hat{e}_i/[\hat{\sigma}_{(i)}\sqrt{(1 - h_{ii})}]$$

where $\hat{\sigma}_{(i)}$ denotes the estimate of σ obtained without the ith observation. As a rough rule, studentized residuals which are larger than about 3 (positive or negative) are 'large' and worthy of further investigation. They may indicate an outlier.

One measure of the **influence** of the ith observation is **Cook's distance** which is given by

$$D_i = r_i^2 h_{ii}/p(1 - h_{ii}).$$

Values of D_i which exceed unity are generally regarded as indicating the presence of an influential observation.

Note that it is possible for an observation with high influence to yield a high (standardized) residual and low leverage (an outlier in regard to the y-value) or a low residual and high leverage (an observation which appears to fit in with the model as fitted to the rest of the data but has a large effect perhaps because the x-values are a 'long way' from the rest of the data. Such an observation may be regarded as an outlier in the x-values.) Thus influential observations may or may not give a y-value which is an outlier, and may or may not cast doubts on the analysis. It is always wise to find out why and how an observation is influential.

A.9 The generalized linear model

Many statistical models can be written in the general form:

observation = systematic component + random component

or in symbols:

$$Y_i = \mu_i + e_i \tag{A.9.1}$$

where Y_i denotes the ith observed random variable, $\mu_i = E(Y_i)$ and $E(e_i) = 0$.

In the general linear model, μ_i is assumed to be a linear function of the explanatory variables, namely

$$\mu_i = \mathbf{x}_i^T \boldsymbol{\beta}$$

where $\mathbf{x}_i$ is the $(p \times 1)$ vector of explanatory variables for the ith observation, and $\boldsymbol{\beta}$ is a $(p \times 1)$ vector of parameters. The errors are assumed to be independent $N(0, \sigma^2)$ variables.

In the **generalized** linear model, the error distribution is allowed to be more general and some function of μ_i is assumed to be a linear combination of the β's. More precisely, the generalized linear model assumes that:

1. The random variables $\{Y_i\}$ are independent and have the same distribution which must be from the exponential family. The exponential family of distributions includes the normal, gamma, exponential, binomial and Poisson distributions as special cases.
2. There is a link function, g (which must be a monotone differentiable function), such that

$$g(\mu_i) = \mathbf{x}_i^T \boldsymbol{\beta} \tag{A.9.2}$$

is a linear function of the x-values. The quantity $\eta_i = \mathbf{x}_i^T \boldsymbol{\beta}$ is sometimes called the **systematic linear predictor**, and then $g(\mu_i) = \eta_i$.

If the Y's and the ε's are normally distributed and g is the identity link function, then it is easy to see that the generalized linear model reduces to the general linear model and so includes ANOVA, regression, etc. However, the generalized model can also describe many other problems.

For example, if the Y's follow a Poisson distribution, and g is the logarithmic function, then we have what is called a **log-linear** model. This is widely applied to count (or frequency) data in contingency tables. If μ_{ij} is the expected frequency in the ith row and jth column, then $(\log \mu_{ij})$ is modelled by the sum of row and column terms (the main effects) and perhaps also by interaction terms. The model can be motivated by noting that $\mu_{ij} = np_{ij}$ where n is total frequency and that, if rows and columns are independent, then $p_{ij} = p_{i.}p_{.j}$ in an obvious notation, so that $\log p_{ij} = \log p_{i.} + \log p_{.j}$ is the sum of row and column effects. The log-linear model emphasizes that it is the *ratio* of frequencies which matters (Exercise G.7(b)).

Another important application is to binary data. Suppose each Y_i follows a binomial distribution with parameters n_i and p_i and that p_i depends on the values of the explanatory variables. Then $\mu_i = n_i p_i$ and there are two link functions in common use. The **logit** transformation of p_i is defined by $\log[p_i/(1 - p_i)]$ and this equals the corresponding link function which is

$$
\begin{aligned}
g(\mu_i) &= \log[\mu_i/(n_i - \mu_i)] \\
&= \log[n_i p_i/(n_i - n_i p_i)] \\
&= \log[p_i/(1 - p_i)].
\end{aligned}
\tag{A.9.3}
$$

If $g(\mu_i) = \mathbf{x}_i^T \boldsymbol{\beta}$ is a linear function of the predictor variables, then the resulting analysis is called **logistic regression** or **logit analysis**. An alternative link function, which often gives results which are numerically very similar, is the **probit** transformation given by

$$
g(\mu_i) = \Phi^{-1}(\mu_i/n_i) = \Phi^{-1}(p_i)
\tag{A.9.4}
$$

where Φ denotes the cdf of the standard normal distribution. $\Phi^{-1}(p_i)$ is usually called the probit of p_i and the resulting analysis is called a **probit analysis**.

To illustrate the use of logistic regression or probit analysis, suppose we observe the proportion of rats dying at different doses of a particular drug (see also Exercise G.5(a)). Suppose that n_i rats receive a particular dose level, say x_i, and that r_i rats die. Then we would expect r_i to follow a binomial distribution with parameters n_i, p_i where p_i is the population proportion which will die at that dose level. Clearly p_i will vary non-linearly with x_i as p_i is bounded between 0 and 1. By taking a logit or probit transform of p_i, we can fit a generalized linear model. For logistic regression we have

$$
\text{logit}(p_i) = \log[p_i/(1 - p_i)] = \alpha + \beta x_i
$$

The median lethal dose, denoted by LD_{50}, is the dose level, x, at which half the rats die (i.e. $p = 0.5$). Then, since $\text{logit}(0.5) = 0$, we find

$$
LD_{50} = -\alpha/\beta
$$

Note the notation ED_{50} is used to denote the median effective dose when the response is not death.

The GLIM package (Appendix B) allows the user to fit four distributions (normal, gamma, Poisson and binomial) and eight link functions in certain combinations which make practical sense. The link functions include square root, exponential and reciprocal transformations as well as the identity, logarithmic, logit and probit transformations.

As in the general linear model, the generalized linear model allows the explanatory variables to be continuous, categorical or indicator-type variables. After fitting the model, the user should look at the standardized residuals (i.e. raw residuals divided by the corresponding standard error) and assess the goodness-of-fit in a somewhat similar way to that used for the general linear model. However, goodness-of-fit is assessed not by looking at sums of squares, but by looking at log-likelihood functions, and ANOVA is replaced by an **analysis of deviance**.

Suppose we are interested in a proposed model with $r(<n)$ parameters. Its log-likelihood is compared with the log-likelihood of the 'full' model, containing n parameters for which $\mu_i = y_i$ for all i. The (scaled) deviance of the model is defined to be twice the difference in log-likelihoods, namely

$$\text{Deviance} = -2 \log\left[\frac{\text{likelihood of proposed model}}{\text{likelihood of full model}}\right] \qquad \text{(A.9.5)}$$

The null model is defined as the model containing no explanatory variables for which μ_i is a constant (so $r = 1$). This model has the largest possible deviance and is often fitted first for comparison purposes.

In order to compare a given model I with a more complicated model II containing m extra parameters, it can be shown that

$$\text{deviance (model I)} - \text{deviance (model II)} \sim \chi_m^2$$

if the model I with fewer parameters is adequate.

For the general linear model, note that the deviance simply reduces to the (scaled) residual sum of squares.

Further reading

Dobson (1990) provides a readable introduction, while McCullagh and Nelder (1989) give a more thorough advanced treatment.

A.10 Sample surveys

Sample surveys are widely used in areas such as market research, sociology, economics and agriculture, to collect both (objective) factual information and (subjective) personal opinions. The idea is to examine a representative subset (called the **sample**) of a specified population.

The (target) **population** is the aggregate of all individuals, households, companies, farms, or of whatever basic unit is being studied, and needs to be carefully defined. The basic units which comprise the population are variously called **sampling units, elementary units, elements** or simply **units**. A complete list of the sampling units is called a **frame**. This list should be as accurate as possible, but may well contain errors so that the sampled population differs somewhat from the target population. For example, electoral registers, which are often used as frames, are always out of date to some extent.

If a survey covers virtually all the units in the population, then we have what is called a complete survey or **census**. However, on grounds of speed, accuracy and cost, a sample is nearly always preferred to a census. If the sampling units are selected by appropriate statistical methods, we have a sample survey.

When the sampling units are human beings, the main methods of collecting information are:

1. face-to-face interviewing;
2. postal surveys;
3. telephone surveys;
4. direct observation.

Face-to-face interviewing is used widely, can give a good response rate, and allows the interviewer some flexibility in asking questions. Field checks are advisable to ensure that interviewers are doing their job properly. Postal surveys are much cheaper to run but generally give a much lower response rate (perhaps as low as 10%). Follow-up requests by post or in person may be necessary. Telephone surveys are increasingly used for many purposes because they are relatively cheap and yield a much higher response rate. However, the possibility of bias must be kept in mind because not everyone has a telephone, and the selection of 'random' telephone numbers is not easy.

A **pilot survey** usually plays a vital role in planning a survey. This is a small-scale version of the survey as originally planned. It is essential for trying out the proposed questions and eliminating teething problems. It should answer the following questions: (i) Is the questionnaire design adequate? (ii) How high is the non-response rate? (iii) How variable is the population? (iv) Should the objectives be changed in any way? The answers to (ii) and (iii) can be useful in determining the sample size required for a given accuracy.

There are many sources of error and bias in sample surveys and it is essential to anticipate them and take precautions to minimize their effect. The simplest type of error, called **sampling error**, arises because a sample is taken rather than a census. Errors of this kind are relatively easy to estimate and control. However, other sources of error, called **non-sampling error**, are potentially more damaging. Possible sources include:

1. the use of an inadequate frame;
2. a poorly designed questionnaire (section A.10.2);
3. interviewer bias;
4. recording and measurement errors;
5. non-response problems (see section A.10.3).

Measurement errors arise, for example, when the respondent gives incorrect answers, either deliberately or unwittingly.

A.10.1 Types of sample design

There are many different types of sample design. The main aim is to select a representative sample, avoid bias and other non-sampling errors, and achieve maximum precision for a given outlay of resources. The two most important types of sampling procedure are random and quota sampling. In **random** sampling, the sample is pre-selected from the entire population using a random selection procedure which gives every member of the population a non-zero, calculable chance of being selected. However, in **quota** sampling, the choice of sampling units is left to the interviewer subject to 'quota controls' designed to ensure that characteristics such as age and social class appear in the sample in a representative way.

A simple random sample involves taking a sample of n units from a population of N units without replacement in such a way that all possible samples size n have an equal chance of being selected. It can then be shown that the sample mean, $\bar{x}$, of a particular variable, x, is an unbiased estimate of the underlying population mean with variance

$$(1 - f)\sigma^2/n$$

where $f = n/N$ is the sampling fraction and σ^2 is the population variance of the x-values. The factor $1 - f$ is called the **finite population correction** to the 'usual' formula for the variance of a sample mean. Simple random samples may be theoretically appealing, but are rarely used in practice for a variety of practical reasons.

In **stratified** random sampling, the population is divided into distinct subgroups, called **strata**, and then a simple random sample is taken from each stratum. The two main reasons for doing this are:

1. to use one's knowledge about the population to make the sample more representative and hence improve the precision of the results;
2. to get information about subgroups of the population when these are of interest in themselves.

If the same sampling fraction is taken from each stratum, we have what is called **proportional allocation**.

Multi-stage sampling arises when the population is regarded as being composed of a number of first-stage units (or primary sampling units), each of which is composed of a number of second-stage units, and so on. A random sample of first-stage units is selected, and then a random sample of second-stage units from selected first-stage units and so on. This type of sample is generally less accurate than a simple random sample of the same size but has two important advantages. First, it permits the concentration of field work by making use of the natural grouping of units at each stage. This can reduce costs considerably and allows a larger sample for a given outlay. Second, it is unnecessary to compile a frame for the entire population. **Cluster sampling** is a special type of multi-stage sampling in which groups or clusters of more than one unit are selected at the final stage and every unit is examined.

Quota sampling does not involve truly random selection and can give biased samples. However, its simplicity means it is widely used. The cost per interview is lower, the sample can be taken more quickly, and no sampling frame is required. However, there is no valid estimate of error and bias may be unwittingly introduced. A widely used rule of thumb is to suppose that the standard error of a value derived from a quota sample is twice as large as the standard error that would result for a simple random sample of the same size.

Two other types of sampling are **judgemental** and **systematic sampling**. In the former, an 'expert' uses his/her knowledge to select a representative sample. This procedure is not random and can be dangerously biased. In systematic sampling, the elements in the sampling frame are numbered from 1 to N. The first unit in the sample is selected at random from the first k units. Thereafter every kth element is selected systematically. The value of k is N/n. This procedure is also not random, but can be very convenient provided that there is no possibility of periodicity in the sampling frame order.

A.10.2 Questionnaire design

The first requirement is to define carefully the objectives of the survey and write down the information that is required. The temptation to include too many questions should be resisted. It is one thing to say that all questions should be clear, concise, simple and unambiguous, but quite another to achieve this in practice. Of course all questions should be rigorously tested and a pilot survey is essential. The design depends to some extent on whether it is to be filled in by the respondent, or by an interviewer, and on whether the data are to be coded at the same time as they are recorded.

Most questions may be classified as **factual** or **opinion**. The latter are harder to construct as they are much more sensitive to small changes in question wording and in the emphasis given by the interviewer.

Most questions may also be classified as **open** (or free-answer) or **closed**. In the latter case, the respondent has to choose from a limited list of possible answers.

The following principles are worth noting in regard to question wording:

1. Use simple, concise everyday language.
2. Make your questions as specific as possible.

3. Avoid ambiguity by trying them out on several different people.
4. Avoid leading questions and the use of unfairly 'loaded' words. In fact it is quite difficult to make questions completely neutral. The use of an implied alternative should be avoided. Thus the last two words of the question 'Do you think this book is well written or not?' are vital as without them the alternative is only implied, and too many people will tend to agree with the interviewer.
5. Do not take anything for granted, as people do not like to admit ignorance of a subject.

A.10.3 The problem of non-response

It is often impossible to get observations from every unit in the selected sample and this gives rise to the problem of **non-response**. This can arise for a variety of different reasons. If non-respondents have different characteristics to the rest of the population, then there is liable to be bias in the results.

In postal surveys, if reminders are sent to people who do not reply to the first letter, then the results from the second wave of replies can be compared with those who replied at once.

In personal interviews, non-response may arise because a respondent refuses to cooperate, is out at the time of call, has moved home, or is unsuitable for interview. There are several methods of coping with 'not-at-homes', including calling back until contact is made, substituting someone else such as a next-door neighbour, subsampling the non-respondents rather than trying to contact them all, and the Politz–Simmons technique in which respondents are asked on how many of the previous five weekday evenings they were at home so that their responses can be weighted accordingly.

A.10.4 Concluding remarks

The usefulness of sample surveys is not in question, but it still pays to view survey results with some scepticism. Different sample designs and slightly different questions may yield results which differ substantially, particularly for sensitive opinion questions. Non-sampling error is generally more important than sampling variation. Indeed, there is a well-known saying in the social sciences that 'any figure which looks interesting is probably wrong!'. While this may be an exaggeration, opinion survey results should be regarded as giving orders of magnitude rather than precise estimates, especially when non-response is present. As one example, a survey of patients at a local hospital produced a response rate of only 30% from a random sample size 500. This low response rate made valid inference about population parameters very difficult. Nevertheless, the reported overwhelming dissatisfaction with one aspect of patient care was enough to justify immediate action.

Further reading

There is much to be said for consulting the classic texts by Cochran (1977) for the theory of sampling, by Moser and Kalton (1971) for many practical details, and by Kish (1965) for a long reference source on both theory and practice. Recent textbooks include Barnett (1991) and Foreman (1991). There are several more specialized books on recent practical developments such as telephone surveys, the use of longitudinal surveys, and the use of consumer panels. Lessler and Kalsbeek (1992) cover the three main sources of non-sampling error, namely (i) the use of an inadequate frame, (ii) non-response and (iii) measurement error.

A.11 The design of experiments

Experiments are carried out in all branches of science. Some are well designed but others are not. This section gives brief guidance on general principles, particularly for avoiding systematic error and increasing the precision of the results. Clinical trials are discussed as a special case in section A.12.

Comparative experiments aim to compare two or more treatments, while in **factorial** experiments the response variable depends on two or more variables or factors. The value that a factor takes in a particular test is called the **level**, and a **treatment combination** is a specific combination of factor levels. The **experimental unit** is the object or experimental material on which a single test, or trial, is carried out.

Before designing an experiment, clarify the objectives carefully and carry out thorough preliminary desk research. Choose the treatments which are to be compared or the factors which are to be assessed. Choose the experimental units which are to be used. Select a suitable response variable which may be a function of the measured variables. Decide how to apply the treatments or treatment combinations to the units, and decide how many observations are needed. If the observations can only be taken one at a time, then the order of the tests must also be decided. If the same experimental unit is used for more than one test, then **carry-over** effects are possible and some sort of **serial** design may be desirable.

One important general principle for eliminating unforeseen bias is that the experimental units should be assigned by a procedure which involves some sort of **randomization**. If the same unit is used for several tests, then the randomization principle should also be applied to the order of the tests. Complete randomization is often impossible or undesirable and some sort of restricted randomization (e.g. using blocking – see below) is often preferred. A second useful principle is that of **replication** in that estimates will be more precise if more observations are taken. Observations repeated under as near identical conditions as possible are particularly helpful for assessing experimental error. A third general principle is *blocking* (see below) whereby any natural grouping of the observations is exploited to improve the precision of comparisons. The **analysis of covariance** (section A.8) is a fourth general approach to improving precision which makes use of information collected on 'concomitant' variables (i.e. variables which 'go together' with the response variable – often measurements made on the experimental units before a treatment is applied).

The simplest type of comparative experiment is the simple (one-way) randomized comparative experiment in which a number of observations are taken randomly on each treatment. The ensuing analysis aims to compare the resulting group means, usually by means of a one-way ANOVA.

The most common type of comparative experiment is the **randomized block** design (Exercise H.2). The tests are divided into groups or blocks of tests which are 'close together' in some way, for example tests made by the same person or tests made on a homogeneous group of experimental units. Then an equal number of observations are taken on each treatment in each block. The order or allocation within blocks is randomized. This design allows between-block variation to be removed so that a more precise comparison of the treatment means can be made. The ensuing analysis usually involves a two-way ANOVA.

There are various more complicated types of comparative experiment such as balanced incomplete block designs and Latin square designs (Exercise E.2), but we do not have space to describe them here.

One important type of factorial experiment is the **complete factorial** experiment in which every possible combination of factor levels is tested the same number of times (e.g. see Exercise E.6). There are various more complicated types of factorial experiment which may involved the idea of **confounding**. Two effects are said to be confounded if it is impossible to distinguish between them (or estimate them separately) on the basis of a given design. In a confounded complete factorial experiment the tests are divided into blocks and it is desirable to arrange the design so that the block effects are confounded with (hopefully unimportant) higher-order interactions. In a fractional factorial experiment, a fraction (e.g. one half or one quarter) of a complete factorial is taken and then every effect is confounded with one or more other effects, which are called **aliases** of one another. The general idea is to choose the design so that it is possible to estimate main effects and important low-order interactions, either uniquely (in confounded complete factorials) or aliased with unimportant higher-order interactions (in fractional factorials). Fractional factorials are involved in the so-called Taguchi methods which aim to minimize performance variation at the product design stage.

Split-plot designs arise when there are blocks, called whole plots, which can be subdivided into subplots. The levels of one factor or treatment are assigned to the whole plots so that the effect of this factor is confounded with the block effect. The levels of the other factors are applied to specified subplots.

A complete factorial experiment is called a **crossed** design since every level of one factor occurs with every level of a second factor. An important alternative class of designs are **nested** (or **hierarchical**) designs. With two factors, denoted by A and B, factor B is nested within factor A if each level of B occurs only with one level of factor A. Then model (A.7.2) would be changed to

$$y_{ijk} = \mu + A_i + B_{j(i)} + \varepsilon_{k(ij)} \qquad \text{(A.11.1)}$$

where, for example, y_{ijk} is the kth observation on response variable with factor A at ith level and factor B at jth level and $B_{j(i)}$ is the effect of factor B at jth level for a specified value of i. (The bracket in the subscript indicates nesting.) Notice that model (A.11.1) contains no terms of the form $\{B_j\}$.

There are a number of specialized designs, such as **composite** designs, which are used in the study of **response surfaces**, where the response variable, y, is an unknown function of several predictor variables and particular interest lies in the maximum (or minimum) value of y.

Optimal designs are concerned with allocating resources so as to achieve the 'best' estimate of the underlying model. 'Best' may be defined in several ways but usually involves the precision of estimated parameters. For example, a D-optimal design is chosen to minimize the determinant of $(X^TX)^{-1}$, in the notation of section A.8, since the variance-covariance matrix of $\hat{\beta}$ in equation (A.8.3) is proportional to $(X^TX)^{-1}$. Although optimal designs are of much theoretical interest, they seem to be little used in practice because the theory requires the precise prior formulation of features (such as the underlying model) which are usually known only partially.

The design and analysis of experiments still relies heavily on the elegance of the solution of the least-squares equations arising from balanced, orthogonal designs. In fact the computing power now available allows most 'reasonable' experiments to be analysed. While it is still desirable for designs to be 'nearly' balanced and 'nearly' orthogonal, the requirement is perhaps not as compelling as it used to be (e.g. Exercise G.3). In a practical situation with special constraints, it is usually possible to use common sense to construct a sensible design with or without reference to standard designs such as those listed in Cochran and Cox (1957).

Further reading

The two classic texts by Cochran and Cox (1957) and Cox (1958) remain useful today. Other texts include Box, Hunter and Hunter (1978) and Montgomery (1991). Steinberg and Hunter (1984) review developments and suggest directions for future research. Hahn (1984) presents some nice examples to illustrate how the general principles of data collection need to be tailored to the particular practical situation.

A.12 Clinical trials

A clinical trial may be described as any form of planned study which involves human beings as medical patients. They are widely used by pharmaceutical companies to develop and test new drugs and are increasingly used by medical researchers to assess a wide variety of medical treatments such as diets, surgical procedures, the use of chemotherapy and different exercise regimes. Tests on healthy human volunteers and on animals have many features in common with clinical trials.

The history of clinical trials is fascinating (see Pocock, 1983, Chapter 2; and section 7.4.2 in Part One). After the thalidomide tragedy in the 1960s, many governments have laid down much stricter regulations for testing new drugs. Partly as a result, clinical trials and toxicological tests now constitute an increasingly important area of experimental design, but are still not very familiar to many statisticians. Thus this section is longer than might be expected! In my limited experience, all experiments on human beings are tricky to handle and doctors have much talent for 'messing up' experiments!

Drug trials are carried out in different phases. For example, there may be preclinical trials on animals, followed by phase I which is concerned with drug safety in human beings. Phase II is a small-scale trial to screen drugs and select only those with genuine potential for helping the patient. Phase III consists of a full-scale evaluation of the drug. Phase IV consists of post-marketing surveillance (or drug monitoring) to check on such matters as long-term side-effects and rare extreme reactions.

Phase III is what many people think of as the clinical trial. There should be a **control** group for comparative purposes which receives the current standard treatment or alternatively a **placebo**. The latter is an inert substance which should have no effect other than that caused by the psychological influence of taking medicine.

The statistician should not be content with simply analysing the results, but rather help in planning the whole trial. It is becoming standard practice to develop a written **protocol** which documents all information about the purpose, design and conduct of the trial. It should describe the type of patient to be studied, the treatments which are to be compared, the sample size for each treatment, the method of assigning treatments to patients (the design), the treatment schedule, the method of evaluating the patient's response to the treatment and the procedure for carrying out interim analyses (if any). The absence of a proper protocol is a recipe for disaster. Deviations from the protocol are in any case likely to occur (e.g. ineligible patients are included), and common-sense decisions have to be made as to what to do about them in order to avoid getting biased results. For example, patient withdrawals (such as taking a patient off a drug due to side-effects) can lead to serious biases if such cases are simply excluded.

The design of the trial is crucial. Randomization should always be involved so that each patient is randomly assigned to the new or standard treatment. There are many potential biases without randomization which may not be apparent at first sight, and non-randomized studies, such as historical retrospective trials, are more likely to give (spurious) significant results. In particular, the systematic alternation of treatments may not give comparable groups, particularly if the doctors know which treatment is being allocated. Many drug trials are carried out in a 'double-blind' way so that neither the patient nor the doctor knows which treatment the patient is receiving. The doctor's judgement of the effect of the treatment is then less likely to be biased. When each patient receives just one treatment, there are various ways of carrying out randomization. Treatments can be allocated randomly over the whole sample or within smaller homogeneous groups (blocks or strata). There are also more specialized designs such as the two-period cross-over trial where each patient receives two treatments, one after the other, with the order being randomized.

The information for each patient is usually recorded on a form. The design of a 'good' form is important. In particular, the data should be suitable for transfer to a computer. It is unwise to try to record too much information.

As results become available, the possibility of carrying out interim analyses and then adjusting the sample size needs to be considered and **sequential** designs are of much current interest. The final analysis should not concentrate too much on testing the effect of the new treatment. It is more important to estimate the size of the effect as well as to assess any side-effects. In publishing results, it is sad that published work hardly ever reports non-significant results or the outcomes of confirmatory studies. It appears that publication is influenced by finding significance. One well-known treatment for cancer was used for a long time because of one historical non-randomized study which gave a significant result even though four subsequent clinical trials showed it to have little or no effect.

Carrying out tests on human beings inevitably raises many ethical questions. One has to balance the welfare of individuals (or animals) in the trial against the potential benefit to the whole population in the future. When do we need to get 'informed consent' from patients? When is randomization ethical? Some doctors are still strongly opposed to randomized trials which take away their right to prescribe a patient the treatment which is believed to be 'best'. However, in seeking to give the 'best' treatment, some doctors will worry that some patients will not get the 'marvellous' new treatment, while others conversely will worry that some patients get the 'dubious' new treatment before it has been fully tested. The issue can only be resolved by a proper randomized trial, which, although difficult and costly, can be very rewarding. While a few treatments (e.g. penicillin) give such striking results that no clinical trial is really necessary, many treatments introduced without proper evaluation are eventually discarded because they are found to have no effect or are even harmful.

Clinical trials are just one way of carrying out research into **epidemiology**, which is the study of health and disease in human populations. Other types of epidemiological study are often observational in nature. For example, cohort studies follow a group of subjects through time to see who develops specific diseases and what risk factors they are exposed to.

Further reading

A good general introductory text for statistics in medical research is by Altman (1991). Pocock (1983) and Friedman, Furberg and Demets (1985) provide excellent introductions to clinical trials. Gore and Altman (1982) discuss the use of statistics in medicine more generally, and there are hopeful signs that sound statistical ideas are starting to permeate the medical profession. For example, Altman *et al.* (1983) lay down guidelines for the use of statistics in medical journals.

A.13 Multivariate analysis

A common objective in multivariate analysis is to simplify and understand a large set of multivariate data. Many techniques are exploratory in that they seek to generate hypotheses rather than test them (section 6.6 of Part One). Some techniques are concerned with relationships between variables, while others are concerned with relationships between individuals or objects. There is also an important distinction between the case where the variables arise on an equal footing and the case where there are response and explanatory variables (as in regression).

This section is primarily concerned with some specific multivariate techniques. However you should begin, as always, by 'looking' at the data. Examine the mean and standard deviation of each variable. Plot scatter diagrams for selected pairs of variables. Look at the matrix of correlations between pairs of variables. If, for example, most of the correlations are close to zero, then there is little linear structure to explain and you may be able to look at the variables one at a time (but look out for non-linear relationships).

Let X denote a p-dimensional random variable with mean vector μ, where $X^T = [X_1, \ldots, X_p]$ and $\mu^T = [E(X_1), \ldots, E(X_p)] = [\mu_1, \ldots, \mu_p]$. The $(p \times p)$ **co-variance** (or **dispersion** or **variance-covariance**) **matrix** of X is given by

$$\Sigma = E[(X - \mu)(X - \mu)^T]$$

so that the (i, j)th element of Σ is the covariance of X_i with X_j. The $(p \times p)$ **correlation matrix** of X, denoted by P, is such that the (i, j)th element measures the correlation between X_i and X_j. Thus the diagonal terms of P are all one. One important special case arises when X has a multivariate normal distribution (already defined in section A.3).

Suppose we have n observations on $\mathbf{X}$. Denote the $(n \times p)$ data matrix by X so that the (i, j)th element x_{ij} is the ith observation on variable j. (Note that vectors are printed in bold type, but matrices are not.) The sample mean vector, $\bar{\mathbf{x}}$, the sample covariance matrix, S, and the sample correlation matrix, R, may be calculated in an obvious way. For example,

$$S = (X - \mathbf{1}\bar{\mathbf{x}}^{\mathrm{T}})^{\mathrm{T}}(X - \mathbf{1}\bar{\mathbf{x}}^{\mathrm{T}})/(n - 1)$$

where $\mathbf{1}$ denotes an $(n \times 1)$ vector of ones.

Principal component analysis is concerned with examining the interdependence of variables arising on an equal footing. The idea is to transform the p observed variables to p new, orthogonal variables, called principal components, which are linear combinations of the original variables ($\mathbf{a}^{\mathrm{T}}\mathbf{X} = \Sigma a_i X_i$) and which are chosen in turn to explain as much of the variation as possible. Thus the first component, $\mathbf{a}_1^{\mathrm{T}}\mathbf{X}$, is chosen to have maximum variance, subject to $\mathbf{a}_1^{\mathrm{T}}\mathbf{a}_1 = 1$, and is often some sort of average of the original variables. Mathematically it turns out that $\mathbf{a}_1$ is the eigenvector of the covariance matrix of $\mathbf{X}$ (or more usually of the correlation matrix) which corresponds to the largest eigenvalue. More generally, the coefficients of the different principal components are the eigenvectors of S (or of R) and the variance of each component is given by the corresponding eigenvalue. Now the sum of the eigenvalues of a square, positive semi-definite real matrix is equal to the sum of the diagonal terms (called the **trace**). Thus for a correlation matrix, whose diagonal terms are all unity, the trace is p so that the proportion of the total variance 'explained' by the first principal component is just λ_1/p, where λ_1 is the largest eigenvalue of R. It is often found that the first two or three components 'explain' most of the variation in the original data, so that the effective dimensionality is much less than p. The analyst then tries to interpret the meaning of these few important components. I have also found it helpful to plot a scatter diagram of the first two components for different individuals in order to try and identify clusters and outliers of individuals. Note that if the X's are linearly dependent (e.g. if say $X_3 = X_1 + X_2$), then S (or R) will be singular and positive semi-definite (rather than positive definite) so that one or more eigenvalues will be zero.

Factor analysis has a similar aim to principal component analysis, namely the reduction of dimensionality. However it is based on a 'proper' statistical model involving $m(<p)$ unobservable underlying factors, $\{f_i\}$, such that

$$X_j = \sum_{i=1}^{m} \lambda_{ji} f_i + e_j.$$

The weights $\{\lambda_{ji}\}$ are called the **factor loadings**. The $\{f_i\}$ are called the **common** factors and the $\{e_j\}$ the **specific** factors. The portion of the variability in X_j explained by the common factors is called the **communality** of the jth variable. The details of model fitting will not be given here. Note that many computer programs allow the user to rotate the factors in order to make them easier to interpret. However, there is a danger that the analyst will try different values of m (the underlying dimensionality) and different rotations until he gets the answer he is looking for! Also note that factor analysis is often confused with principal component analysis, particularly as the latter is sometimes used to provide starting values in factor analysis model building.

In my experience, social scientists often ask for help in carrying out a factor analysis, even though they have not looked at the correlation matrix and do not really understand what is involved. If most of the correlations are 'small', then the variables are essentially independent and so there is no point in carrying out a factor analysis (or a principal component analysis where the eigenvalues are likely to be nearly equal). If, on the other hand, all the correlations are 'large', then I would be suspicious that the variables are all measuring the same thing in slightly different ways. (This often happens in psychology with attitude questions.) Yet another possibility is that the variables split into groups such that variables within a group are highly correlated but variables in different groups are not. Such a description may be a better way of understanding the data than a factor analysis model. Only if the correlation matrix contains high, medium and low correlations with no discernible pattern would I consider carrying out a factor analysis (and perhaps not even then!).

Multi-dimensional scaling is concerned with the relationship between individuals. The idea is to produce a 'map', usually in two dimensions, of a set of individuals given some measure of similarity or dissimilarity between each pair of individuals. **Classical scaling** is appropriate when the dissimilarities are approximately Euclidean distances. Then there is a duality with principal component analysis in that classical scaling is essentially an eigenvector analysis of XX^T (ignoring mean-correction terms) whereas principal component analysis is an eigenvector analysis of X^TX. **Ordinal scaling**, or non-metric multi-dimensional scaling, only uses the ordinal properties of the dissimilarities and involves an iterative numerical procedure. Having produced a map, one aim is to spot clusters and/or outliers (as when plotting the first two principal components, but using a completely different type of data).

Cluster analysis aims to partition a group of individuals into groups or clusters which are in some sense 'close together'. There is a wide variety of procedures which depend on different criteria, on different numerical algorithms and on different objectives. For example, some methods allocate individuals to a prescribed number of clusters, while others allow the number of clusters to be determined by the data. Other procedures aim to find the complete hierarchical structure of the data in a hierarchical tree or dendrogram. As a further source of confusion, I note that cluster analysis is variously called classification and taxonomy. In my experience the clusters you get depend to a large extent on the method adopted and the criteria employed.

Correspondence analysis is primarily a technique for displaying the rows and columns of a two-way contingency table as points in dual low-dimensional vector spaces, but may be extended to other non-negative data matrices of a suitable form. Given an $(n \times p)$ non-negative data matrix, find the row and column sums and let R be the $(n \times n)$ diagonal matrix of row sums and C the $(p \times p)$ diagonal matrix of column sums. Then $R^{-1}X$ and $C^{-1}X^T$ are the row-profile and column-profile matrices where we simply divide each observation by the appropriate row or column sum. The analysis then essentially consists of an eigenvector analysis of $(R^{-1}X)(C^{-1}X^T)$ or $(C^{-1}X^T)(R^{-1}X)$, whichever is the smaller matrix.

There are many techniques based on the assumption that $\mathbf{X}$ has a multivariate normal distribution which are often natural generalizations of univariate methods based on the normal distribution. For example, to compare a sample mean vector $\bar{\mathbf{x}}$ with a population mean vector $\boldsymbol{\mu}_0$, the test statistic, called Hotelling's T^2, is given by $n(\bar{\mathbf{x}} - \boldsymbol{\mu}_0)^T S^{-1}(\bar{\mathbf{x}} - \boldsymbol{\mu}_0)$ and is the 'natural' generalization of the square of the univariate t-statistic. The **multivariate analysis of variance** (MANOVA) is the natural extension of ANOVA. The problem then consists of comparing matrices containing sums of squares and cross-products by 'reducing' each matrix to a single number such as its determinant or its trace. I have never found this appealing. When MANOVA rejects a null hypothesis, a technique called **canonical variates analysis** can be used to choose those linear compounds of the form $\mathbf{a}^T\mathbf{X}$ which best show up departures from the null hypothesis when a univariate ANOVA is carried out on the observed values of $\mathbf{a}^T\mathbf{X}$. **Discriminant analysis** is concerned with finding the 'best' linear compound $\mathbf{a}^T\mathbf{X}$ for distinguishing between two populations. The **multivariate analysis of covariance** (MANOCOVA) is the multivariate generalization of the (univariate) analysis of covariance. The method of **canonical correlations** is used to examine the dependence between two sets of variables, say $\mathbf{X}_1$ and $\mathbf{X}_2$. Let $U = \mathbf{a}^T\mathbf{X}_1$ and $V = \mathbf{b}^T\mathbf{X}_2$ be linear compounds of $\mathbf{X}_1$ and $\mathbf{X}_2$ and let ρ denote the correlation between them. Then, for example, the first pair of canonical variates, say $\mathbf{a}_1^T\mathbf{X}_1$ and $\mathbf{b}_1^T\mathbf{X}_2$ are chosen so as to maximize ρ and the resulting value is called the (first) canonical correlation. Further canonical variates can be found, orthogonal to each other, in a similar way. Canonical correlation has never found much support among users because the highest correlations can relate uninteresting linear compounds which make little contribution to the total variability of the data. Several new methods are now being investigated to study the relationships between two data matrices.

One important idea which arises, directly or indirectly, in many multivariate techniques is the use of the **singular value decomposition**. This says that if A is an $(n \times p)$ matrix of rank r, then A can be written as

$$A = ULV^T$$

where U, V are matrices of orders $n \times r$ and $p \times r$, respectively, which are column orthonormal (i.e. $U^TU = V^TV = I_r$) and L is an $(r \times r)$ diagonal matrix with positive elements. In particular, if A is the $(n \times p)$ data matrix, with (i, j)th element x_{ij}, then $\sum_{i,j} x_{ij}^2$ equals the sum of the squared diagonal elements of L.

If the latter are arranged in descending order of magnitude, it may be possible to approximate the variation in the data with the two or three largest elements of L together with the corresponding rows of U and V. If instead A is a (square, symmetric) variance-covariance matrix, then U and V are identical and are formed from the eigenvectors of A, while the diagonal elements of L are the eigenvalues of A. This decomposition is used in principal component analysis.

Further reading

Three general introductory textbooks are by Chatfield and Collins (1980), Krzanowski (1988) and Mardia, Kent and Bibby (1979). A short, readable non-mathematical introduction is given by Manly (1986). The book by Greenacre (1984) is recommended for its treatment of correspondence analysis.

A.14 Time-series analysis

A time series is a set of observations made sequentially through time. A time series is said to be **continuous** when observations are taken continuously through time, but is said to be **discrete** when observations are taken at discrete times, usually equally spaced, even when the measured variable is continuous. A continuous series can always be sampled at equal intervals to give a discrete series which is the main topic of this section.

The special feature of time-series analysis is that successive observations are usually *not* independent and so the analysis must take account of the order of the observations. The main possible objectives are (i) to describe the data, (ii) to find a suitable model and (iii) to forecast future values and/or control future behaviour of the series. If future values can be predicted exactly from past values, then the series is said to be **deterministic**. However, most series are **stochastic** in that the future is only partly determined by past values.

The first step in the analysis is to construct a **time plot** of each series. Features such as trend (long-term changes in the mean), seasonal variation, outliers, smooth changes in structure and sudden discontinuities will usually be evident (Exercises F.1–F.3). Simple descriptive statistics may also be calculated to help in summarizing the data and in model formulation.

An observed time series may be regarded as a realization from a **stochastic process**, which is a family of random variables indexed over time, denoted by $\{X_t\}$ in discrete time. A **stationary** process has constant mean and variance and its other properties also do not change with time. In particular, the **autocovariance** of X_t and X_{t+k}, given by

$$\gamma_k = E[(X_t - \mu)(X_{t+k} - \mu)] \qquad (\text{A}.14.1)$$

where $\mu = E(X_t)$, depends only on the time lag, k, between X_t and X_{t+k}. The **autocorrelation coefficient** at lag k is given by $\rho_k = \gamma_k/\gamma_0$ and the set of coefficients $\{\rho_k\}$ is called the **autocorrelation function** (acf). The **spectrum** of a stationary process is the discrete Fourier transform of $\{\gamma_k\}$, namely

$$f(\omega) = \frac{1}{\pi}\left(\gamma_0 + 2\sum_{k=1}^{\infty}\gamma_k \cos \omega k\right) \tag{A.14.2}$$

where ω denotes the angular frequency such that $0 \le \omega \le \pi$. Note that this can be written in several equivalent ways, and that γ_k is the inverse Fourier transform of $f(\omega)$. For multivariate time series, there is interest in relationships between series as well as within series. For example the **cross-correlation function** of two stationary series $\{X_t\}$ and $\{Y_t\}$ is a function $\rho_{XY}(k)$ which measures the correlation between X_t and Y_{t+k}. The **cross-spectrum** is the discrete Fourier transform of $\rho_{XY}(k)$.

There are many useful classes of model, both stationary and non-stationary. The simplest model, used as a building brick in many other models, is the **purely random process**, or **white noise**, which is henceforth denoted by $\{Z_t\}$ and defined as a sequence of independent, identically distributed random variables with zero mean and constant variance. The acf of $\{Z_t\}$ is given by

$$\rho_k = \begin{cases} 1 & k = 0 \\ 0 & \text{otherwise} \end{cases}$$

which corresponds to a constant spectrum.

An **autoregressive** process of order p (AR(p)) is defined by

$$X_t = \varphi_1 X_{t-1} + \cdots + \varphi_p X_{t-p} + Z_t \tag{A.14.3}$$

where $\varphi_1, \ldots, \varphi_P$ are constants. A **moving average** process of order q (MA(q)) is defined by

$$X_t = Z_t - \theta_1 Z_{t-1} - \cdots - \theta_q Z_{t-q} \tag{A.14.4}$$

where $\theta_1, \ldots, \theta_q$ are constants. By using the backward shift operator, B, for which $BX_t = X_{t-1}$, we may combine (A.14.3) and (A.14.4) to give a mixed **autoregressive-moving average** process (ARMA(p, q)) in the form

$$\varphi(B)X_t = \theta(B)Z_t \tag{A.14.5}$$

where $\varphi(B) = 1 - \varphi_1 B - \cdots - \varphi_p B^p$ and $\theta(B) = 1 - \theta_1 B - \cdots - \theta_q B^q$ are polynomials in B of order p, q respectively. Model (A.14.5) is stationary provided that the roots of $\varphi(B) = 0$ lie outside the unit circle and is said to be **invertible** if the roots of $\theta(B) = 0$ lie outside the unit circle. Invertibility ensures that there is a unique MA process for a given acf. Non-stationary processes can often be made stationary by taking differences such as $W_t = (1 - B)^d X_t$. On substituting W_t for X_t in (A.14.5) we have an **integrated ARMA** model of order p, d, q denoted by ARIMA(p, d, q). In particular, the random walk model defined by

$$X_t = X_{t-1} + Z_t \qquad (A.14.6)$$

is non-stationary and can be regarded as an ARIMA $(0, 1, 0)$ model. It is useful for approximating some economic time series such as exchange rates and share prices. A more general seasonal ARIMA model can also be defined to cope with series showing seasonal variation, by including appropriate polynomials in B^s in equation (A.14.5) where s is the number of observations per season. A **multivariate ARIMA** (or VARIMA) model can also be defined as in equation (A.14.5) with X_t, Z_t vectors and $\varphi(B)$, $\theta(B)$ matrix polynomials. In particular, if the matrices are upper-triangular (with zeros below the diagonal, so that there is no feedback between 'inputs' and 'outputs'), then we have what is called a **transfer-function model**. ARIMA models are useful for describing many, but not all, observed time series.

Another important class of models is the traditional trend-and-seasonal model. In describing trend, it is important to distinguish between a **global** linear trend (e.g. level at time $t = \mu_t = a + bt$), a **local** linear trend (e.g. $\mu_t = a_t + b_t t$, where the slope b_t and intercept a_t changes slowly with time), and a non-linear trend. In particular, series showing exponential growth need careful handling, perhaps by taking logarithms. In describing seasonality, one must distinguish between additive seasonality and multiplicative seasonality (where the seasonal effects are proportional to the local mean). The seasonal factors may be held constant or updated. They are usually normalized to sum to zero in the additive case and to average to one in the multiplicative case.

Another broad class of models, arousing much current interest, is the class of **state space models** for which optimal forecasts may be computed using a recursive estimation procedure called the **Kalman filter**. The latter is widely used in control engineering. Unfortunately, there is no standard notation. The simple univariate state-space model considered here assumes that the observation Y_t at time t is given by

$$Y_t = \mathbf{h}_t^{\mathrm{T}} \boldsymbol{\theta}_t + n_t \qquad (A.14.7)$$

where $\boldsymbol{\theta}_t$ denotes what is called the **state vector**, which describes the 'state of nature' at time t, $\mathbf{h}_t$ denotes a known vector and n_t denotes the observation error. The state vector cannot be observed directly (i.e. is unobservable) but is known to be updated by the equation

$$\boldsymbol{\theta}_t = G_t \theta_{t-1} + \mathbf{w}_t \qquad (A.14.8)$$

where the matrix G_t is assumed known and $\mathbf{w}_t$ denotes a vector of deviations. Equation (A.14.7) is called the **observation** (or **measurement**) equation, while (A.14.8) is called the **transition** (or **system**) equation. The error, n_t, is assumed to be $N(0, \sigma_n^2)$, while $\mathbf{w}_t$ is assumed to be multivariate normal with zero mean and known variance-covariance matrix W_t and to be independent of n_t. One simple example is the **linear growth model**, sometimes called a **structural trend model**, for which

$$Y_t = \mu_t + n_t$$

where
$$\mu_t = \mu_{t-1} + \beta_{t-1} + w_{1t}$$

and
$$\beta_t = \beta_{t-1} + w_{2t}.$$

(A.14.9)

Here the state vector $\theta_t^T = (\mu_t, \beta_t)$ consists of the local level, μ_t, and the local trend, β_t, even though the latter does not appear in the observation equation. We find $\mathbf{h}_t^T = (1, 0)$ and

$$G_t = \begin{bmatrix} 1 & 1 \\ 0 & 1 \end{bmatrix}$$

are both constant through time.

The state-space model can readily be generalized to the case where Y_t is a vector, and many standard time-series models, such as regression, ARIMA and structural models, can be put into this formulation.

Let $\hat{\theta}_{t-1}$ denote the minimum mean square error estimator of θ_{t-1} based on information up to and including Y_{t-1}, with variance-covariance matrix P_{t-1}. The Kalman filtering updating procedure has two stages which may be derived via least-squares theory or using a Bayesian approach. The prediction equations (stage I) estimate θ_t at time $(t-1)$ in an obvious notation by

$$\hat{\theta}_{t|t-1} = G_t \hat{\theta}_{t-1}$$

with

$$P_{t|t-1} = G_t P_{t-1} G_t^T + W_t.$$

When Y_t becomes available, the prediction error is

$$e_t = Y_t - \mathbf{h}_t^T G_t \hat{\theta}_{t-1}$$

and the stage II updating equations are:

$$\hat{\theta}_t = \hat{\theta}_{t|t-1} + K_t e_t$$

and

$$P_t = P_{t|t-1} - K_t \mathbf{h}_t^T P_{t|t-1}$$

where

$$K_t = P_{t|t-1} \mathbf{h}_t / (\mathbf{h}_t^T P_{t|t-1} \mathbf{h}_t + \sigma_n^2)$$

is called the **Kalman gain matrix**, which in the univariate case is just a vector.

The choice of an appropriate model depends on prior information, the objectives, the initial examination of the time plot and an assessment of various more complicated statistics such as autocorrelations. The **sample autocorrelation function** (acf) or **correlogram** is defined by

$$r_k = c_k/c_0 \qquad \text{for } k = 0, 1, 2, \ldots$$

where

$$c_k = \sum_{t=1}^{n-k} (x_t - \bar{x})(x_{t+k} - \bar{x})/n$$

Roughly speaking, values of r_k which exceed $2/\sqrt{n}$ in absolute magnitude are significantly different from zero. The **Durbin–Watson test** essentially looks at $2(1 - r_1)$ and so has expected value 2 for random series. The **partial acf** is another useful diagnostic tool. The partial autocorrelation at lag k is the correlation between X_t and X_{t+k} in excess of that already explained by autocorrelations at lower lags. An analysis based primarily on the correlogram is sometimes called an **analysis in the time domain**. In particular the Box–Jenkins approach, based on fitting ARIMA models, involves a three-stage model-building procedure, namely (a) identifying a suitable ARIMA model (by looking at the correlogram and other diagnostic tools), (b) estimating the model parameters, and (c) checking the adequacy of the model (primarily by looking at the one-step-ahead errors).

An **analysis in the frequency domain** is based primarily on the sample spectrum which can be obtained either by taking a truncated weighted Fourier transform of the acf or by smoothing a function called the **periodogram** which is obtained from a Fourier analysis of the observed time series. For long series the fast Fourier transform can be used to speed calculations.

Note that the correlogram and sample spectrum can be tricky to interpret, even by an 'expert'. Multivariate diagnostics are even harder to interpret.

Before carrying out a time-series analysis, it may be advisable to modify the data either by applying a power transformation (e.g. $Y_t = \log(X_t)$) or by applying a linear digital filter (e.g. $Y_t = \sum_j c_j X_{t-j}$ where the $\{c_j\}$ are the filter weights). In particular, a variety of filters are available for detrending or deseasonalizing time series. Note that **low-pass** filters are filters which remove high-frequency variation, while **high-pass** filters remove low-frequency variation. First-order differencing, for example, can be regarded as a filter with $c_0 = 1$ and $c_1 = -1$. This filter removes trend and is of high-pass form.

Our final topic is **forecasting**. There are many different procedures available which may be categorized as **univariate** (or projection) or **multivariate** (or causal) or judgemental. A univariate forecast is based only on the present and past values of the time series to be forecasted. Forecasts may also be categorized as **automatic** or **non-automatic**. Most univariate forecasting procedures can be put into automatic mode, as may be necessary when forecasting large numbers of series in stock control. The choice of method, and hence of underlying model, depends on a variety of practical considerations including objectives, prior information, the properties of the given data as revealed by the time plot(s), the number of observations available, and so on. **Exponential smoothing** is one simple widely-used projection method. The forecast one-step ahead made at time n is denoted by $\hat{X}(n, 1)$ and is a geometric sum of past observations, namely

$$\hat{X}(n, 1) = \alpha X_n + \alpha(1 - \alpha)X_{n-1} + \alpha(1 - \alpha)^2 X_{n-2} + \cdots$$

This can be rewritten in a more useful updating form as

$$\hat{X}(n, 1) = \alpha X_n + (1 - \alpha)\hat{X}(n - 1, 1) \qquad \text{recurrence form, or}$$

$$\hat{X}(n, 1) = \hat{X}(n - 1, 1) + \alpha e_n \qquad \text{error-correction form}$$

where $e_n = X_n - \hat{X}(n - 1, 1)$ is the one-step-ahead forecasting error. In Winters (or Holt–Winters) forecasting, the local level, local trend and local seasonal factor are all updated by exponential smoothing. Variants of exponential smoothing are used more in practice than other methods such as Box–Jenkins and state-space forecasting. Multivariate forecasting methods, based on multivariate ARIMA, regression or econometric models are much more difficult to handle than univariate methods and do not necessarily give better forecasts for a variety of reasons (e.g. the underlying model changes or the predictor variables may themselves have to be forecasted). Of course forecasts involve extrapolation and should be regarded as conditional statements about the future assuming that past trends continue. Hence the well-known definition: 'Forecasting is the art of saying what will happen and then explaining why it didn't'! It is salutary to end with the following rhyme:

A trend is a trend is a trend.
The question is, will it bend?
Will it alter its course
Through some unforeseen cause
And come to a premature end?

– Alex Cairncross

Further reading

Chatfield (1989), Kendall and Ord (1990) and Wei (1990) are general introductory books. Granger and Newbold (1986) focuses on forecasting. Priestley (1981) covers more advanced topics such as multivariate processes, spectral analysis, and control.

A.15 Quality control and reliability

These two topics are an important part of industrial statistics. Statistical **quality control** is concerned with controlling the quality of a manufactured product using a variety of statistical tools such as the Shewhart control chart. One important class of problems is **acceptance sampling**, which is concerned with monitoring the quality of manufactured items supplied by a manufacturer to a consumer in batches. The problem is to decide whether the batch should be accepted or rejected on the basis of a sample randomly drawn from the batch. A variety of sampling schemes exist. If the items in a sample are classed simply as 'good' or 'defective', then we have what is called **sampling by attributes**. However, if a quantitative measurement (such as weight or strength) is taken on each item, then we have what is called **sampling by variables**. In a single sampling attributes plan, a sample size n is taken and the batch is accepted if the number of defectives in the sample is less than or equal to an integer c called the **acceptance number**. Double sampling is a two-stage extension, while in **sequential** sampling a decision is taken after each observation as to whether to accept or reject or continue sampling.

The performance of an attributes sampling scheme may be described by the **operating characteristic** (or OC) curve which plots the probability of accepting a batch against the proportion of defectives in the batch. The proportion of defectives in a batch which is acceptable to the consumer is called the **acceptable quality level** (AQL). The probability of rejecting a batch at this quality level is called the **producer's risk**. The percentage of defectives in a batch which is judged to be unacceptable to the consumer is called the **lot tolerance percent defective** (LTPD). The probability of accepting a batch at this quality level is called the **consumer's risk**.

Some sampling schemes allow rejected batches to be subject to 100% inspection and rectification. Then the **average outgoing quality** (AOQ) for a particular underlying value of p is the overall average proportion of defectives in batches actually received by the consumer. The **AOQ limit** (AOQL) is the highest (worst) value of the AOQ.

Some schemes allow for different levels of inspection (e.g. normal, tightened or reduced) according to the recent quality observed.

A second important branch of statistical quality control is concerned with **process control**. The problem is to keep a manufacturing process at a specified stable level. Samples are taken regularly so as to detect changes in performance. The causes of these changes should then be found so that appropriate corrective action can be taken. The most commonly used tool is the (Shewhart) **control chart**, on which a variable, which is characteristic of the process quality, is plotted against time. If the observed variable has target value T and residual standard deviation σ when the process is under control, then the graph may have warning lines inserted at $T \pm 2\sigma$ and action lines at $T \pm 3\sigma$. Rather than plot every single observation, it is often convenient to plot the results from regular small samples of the same size n (usually between about 4 and 20). The average quality can be checked by plotting successive sample means on a control chart, called an $\bar{X}$-chart. This will have action lines at $T \pm 3\sigma/\sqrt{n}$. The quality variability can also be checked by plotting successive sample ranges on a control chart, called an *R*-chart. An alternative type of control chart is the **cumulative sum** or **cusum** chart. If x_t denotes the process quality characteristic at time t, then the cumulative sum of deviations about the target, T, is given by

$$S_t = \sum_{k \le t} (x_k - T)$$

and S_t is plotted against t. The local mean of the process corresponds to the local gradient of the graph, which should be about zero when the process is under control. There are also special control charts for sampling by attributes, such as the *p*-chart which plots the proportion defective in successive samples.

As in most areas of statistics, basic theory needs to be supplemented by an understanding of practical problems. For example, in acceptance sampling, batches are often **not** homogeneous and it may be difficult to get random samples. Some checks on quality inspectors may also be desirable as in my experience inspection standards may vary considerably, either because of negligence or because the inspectors genuinely apply different standards as to what constitutes a defective.

There has recently been a change in emphasis regarding quality control with the realization that statistical methods alone will not solve quality problems. Rather there has to be a revolution in the way that systems are managed. The success of Japanese industry during the 1970s and 1980s depended to a large extent on the successful implementation of a complete quality control package, often called **Total Quality Management** (TQM). This approach is now permeating business and commerce throughout North America and Europe, inspired particularly by W. E. Deming. There may be rather little that is new in terms of statistics, but there is a complete change in the overall attitude to quality and a combination of many different ideas into a coherent working package. A first obvious point is that good quality pays! It reduces manufacturing costs, increases customer satisfaction, reduces claims under warranty and can give better product specification. A second point is that the management of a system is crucial. Traditional management has a hierarchical, authoritarian structure. In TQM, a system (whether it is a factory, a company, a hospital or whatever) is seen as a network of interdependent components which should work together to accomplish the aims of the system (which must obviously be clearly stated). Everyone needs to work together as a team to improve the system. When things go wrong, it is usually the fault of the system and not the people, and the latter must know how they contribute to the system. As a trivial example, a person washing a table must know if it is to be used as a desk or as an operating table! Thus we need to combine statistical methods with good human relations and good organizational practice. Some general advice is:

1. Get manager and workers working together and create an environment in which people are not afraid to tell their superiors about faults or problems (avoid management by fear); Japanese authors advocate the use of **quality circles** in which all levels of employees get together to discuss improvements.

2. Avoid confrontation and competition between system components. If you make your suppliers, customers or employees into losers, then the system will suffer as a result.
3. Constant quality improvement should be a way of life (the only acceptable defect rate is zero!).
4. The statistician should work as a quality detective rather than as a policeman. Prevention is better than detection/correction (as in health and crime).
5. Prefer simple statistical tools which will be intelligible to engineers and technicians, such as Ishikawa's seven tools (which include the histogram, the scatterplot and check sheets).
6. Design quality into the product. Taguchi methods comprise an important modern approach wherein experiments are carried out at the product design stage. They aim to optimize some quality characteristic at the same time as minimizing system variability and making the system insensitive to environmental changes. They often use fractional factorial designs. Some statistical aspects of the approach are controversial, but they have succeeded in capturing the imagination of managers and have done much to improve quality by emphasizing the importance of good design.

The topic of **reliability** is of vital importance in manufacturing industry. The reliability of an item is a measure of its quality and may be defined in various ways. It is usually the probability that it will function successfully for a given length of time under specified conditions. One way of measuring reliability is to test a batch of items over an extended period and to note the failure times. This process is called **life testing**. The distribution of failure times may then be assessed, and in particular the mean life and the proportion failing during the warranty period may be estimated. Many similar problems arise in epidemiology in medical experiments on animals or humans where survival analysis examines lifetimes for different medical conditions and/or different drug treatments.

Suppose an item begins to function at time $t = 0$, and let the random variable, T, denote the lifetime (or survival time or time to failure). The cdf and pdf of T will be denoted by $F(t)$, $f(t)$ respectively. The function $R(t) = 1 - F(t) = \text{Prob}(T > t)$ is called the **reliability** function (or **survivor** function). The function $h(t) = f(t)/R(t)$ is called the **hazard** function (or **conditional failure rate** function since $h(t)\Delta t$ is the conditional probability of failure in the interval t to $t + \Delta t$ given that the item has survived until time t). The four functions are equivalent complementary ways of describing the probability distribution of failure times (i.e. the distribution of T). There are various probability distributions which may or may not be suitable to describe the distribution of T. In the notation of table A.3.1, a simple model is that T has an exponential distribution in which case it can be shown that $h(t)$ is a constant, λ. A more realistic assumption is that $h(t)$ increases with time and then the Weibull distribution may be appropriate with $m > 1$, since it can be shown that $h(t) = m\lambda t^{m-1}$. The normal and log-normal distributions are also sometimes used.

There are many types of life test. For example, N items may be tested, with or without replacement, until a specified number of failures has occurred (**sample-truncated**) or until a specified time has elapsed (**time-truncated**). Some sort of sequential procedure may be sensible. Failure times are rarely in-dependent. There may be initial failures due to 'teething' problems or wear-out failures due to fatigue. If some items do not fail before the end of the test, then the resulting lifetimes are said to be **censored** (on the right) and they may complicate the analysis (Exercises B.5 and E.4). The empirical reliability function (i.e. the observed proportion lasting to at least time t) is useful to plot the data graphically but may need to be modified with censored data or when there may be failures for more than one reason (a competing risk situation). The **Kaplan–Meier estimate** of the reliability function is a non-parametric estimate which takes account of censored observations in a 'obvious' way, namely, if d_j deaths occur at time t_j (where $t_1 < t_2 < \cdots$) and there are n_j subjects at risk at t_j, then $\hat{R}(t) = \prod_{t_j < t}(n_j - d_j)/n_j$. There are various ways of plotting empirical functions, such as the reliability function or hazard function, on various types of probability paper. For example, Weibull paper is designed so that the empirical cdf is approximately linear if the underlying distribution is Weibull.

A more sophisticated analysis may try to relate the reliability or hazard function to the values of one or more explanatory variables. For example, the **log-linear proportional hazards model**, or **Cox's model**, assumes that

$$h(t) = h_0(t)\exp(\boldsymbol{\beta}^{\mathrm{T}}\mathbf{x})$$

where $h_0(t)$ is a baseline hazard function under standard conditions, $\mathbf{x}$ is the vector of explanatory variables and $\boldsymbol{\beta}$ is a parameter vector to be estimated from the data.

A different type of reliability problem is to estimate the reliability of a system given the reliabilities of the components which make up the system. For example, the reliability of a space rocket (e.g. the probability that it completes a mission successfully) has to be estimated from the reliabilities of the individual components. A set of components is said to be connected **in series** if the failure of any component causes failure of the system. A set of components is connected **in parallel** if the system works provided that at least one of the components works. Redundant components are often connected to a system in parallel in order to improve the reliability of the system. It is relatively easy to evaluate a system's reliability when its components function independently, but un-fortunately this is often not the case.

Further reading

There are many books on quality control, such as Duncan (1974) and Bissell (1994). A non-technical account of the Japanese approach to quality management is given by Ishikawa (1985). An influential message about the need for radical change is given by Deming (1986) in an idiosyncratic, but highly effective, book containing numerous cautionary tales. Deming proposes 14 guidelines for top management and they are listed by Barnard (1986) in an entertaining paper on the history of industrial statistics and methods for improving the quality performance of manufacturing industry. Deming's guidelines include such maxims as 'Drive out fear, so that everyone may work effectively for the company' and 'Institute a vigorous program of education and retraining'. There are several good books on reliability, life testing and survival modelling, including those by Crowder *et al.* (1991) and Collett (1994).

Tailpiece

I am an observation.
I was captured in the field.
My conscience said 'co-operate'
My instinct said 'don't yield'.
But I yielded up my data
Now behold my sorry plight
I'm part of a statistic
Which is not a pretty sight.
The Bootstrap and the Jackknife
Oh, the tortures I've endured
They analyse my variance
Until my meaning is obscured.
But I've a plan to beat them
I'll climb up in the trees
Pretend I am a chi-square
And get freedom by degrees.

– after T.P.L.

Appendix B
MINITAB and GLIM

This appendix gives brief reference notes on two important packages, called MINITAB and GLIM. Full details may be found in the appropriate reference manuals. Note that both packages will be updated in due course, in which case some commands may change. Of course many other packages could have been included, but the author has concentrated on two packages he is familiar with.

B.1 Minitab

MINITAB is a general-purpose, interactive statistical computing system which is very easy to use and which is widely used in teaching. These notes refer to Release 9.1 (see also the helpful book by Ryan and Joiner, 1994, and the various MINITAB manuals). Log in to your computer and go into the MINITAB package.

Data

Data and other numbers are stored in columns, denoted by $c1, c2, \ldots, c1000$, in a worksheet consisting of not more than 1000 columns. It is also possible to store up to 100 matrices, denoted by $m1, m2, \ldots, m100$, and up to 1000 constants, denoted by $k1, \ldots, k997$ ($k998$–$k1000$ store special numbers). The total worksheet size available depends on the computer, but is usually 'large'.

Commands

When you want to analyse data, you type the appropriate commands. There are commands to read, edit and print data, to manipulate the columns of data and do arithmetic, to plot the data, and to carry out various statistical analyses such as regression, t-tests and ANOVA. Recent releases incorporate more advanced procedures, such as factor analysis, as well as providing better graphics.

Help

The command HELP HELP gives information about help commands. HELP OVERVIEW gives general help. A command such as HELP SET gives help on a particular command such as SET.

Prompts

Before entering a command, wait for the computer to prompt you with MTB>. When entering data with the SET, READ, or INSERT commands, the prompt is DATA>. When a command ends with a semi-colon, the prompt for a subcommand is SUBC>.

Examples

1. Suppose you want to add some lengths held in c1 to the corresponding breadths held in c2 and put the sums into a new column, say c3. The command is simply: LETΔc3 = c1 + c2 (where Δ denotes a space). For example, if c1 = (1, 2), c2 = (5, 9), then c3 = (6, 11).
2. LET c4 = 2*c1
   Here, each element of c1 is multiplied by 2 and the results are put into c4.
3. (a) MEAN c1
      This finds the average of the values in c1 and prints it.
   (b) MEAN c1, k2
      Finds the average of the values in c1 and puts it into k2.
   (c) MEAN c1 [, k2]
      The square brackets indicate that what is inside is optional. The brackets themselves are never typed.

Termination of lines

All lines are terminated and sent to the computer by the Return key.

Error messages

If a command is not properly formulated, you will be so informed.

Finishing

When you have finished using Minitab, type STOP. This command exits you from Minitab.

More on commands

There are about 150 recognized commands and only an important subset will be reviewed here. The computer accepts commands of various lengths but only looks at the first four letters (or fewer if the command is less than four letters long such as LET). Thus the commands HISTOGRAM and HIST are equivalent. Commands may be in upper or lower case or a mixture (e.g. LET or let or Let). If you ever get the prompt CONTINUE?, say YES. Many commands have subcommands to increase their versatility. Then the main command must end in a semicolon and the subcommand in a full stop.

Data entry

1. Data can be entered using the SET command. Here is an example:
   MTB> SET c1
   DATA> 2.6 3.9 4.9 5.7
   DATA> 1.7 2.8 5.2 3.9
   DATA> END
   This specifies the first eight elements of c1. Numbers are entered in free format separated by a blank(s) or a comma or both. Do not try to get more than about 60 characters on a line. It is better to put the same number of numbers on each line in the same format so as to help check the input. There are various shortcut tricks such as:

   MTB> SET c14 ⎫
   DATA> 12:48   ⎬ Puts consecutive integers from
   DATA> END   ⎭ 12 to 48 into c14.

2. To enter a data matrix, each column (or row) can be entered separately using the SET command. Alternatively, one could use the READ command, e.g. READ c1 c2 c3 – and then each row of data should contain three numbers, one for each column. Alternatively you can use e.g. READ 8 3 m8, to read in a (8 × 3) matrix in eight rows of three elements into matrix m8.

3. SET 'filename' c1 – reads a file named filename into c1. Note the quotation marks.

Editing and manipulating data

If you make a mistake in entering data, or wish to manipulate the data in some way, there are various useful commands such as LET, INSERT, DELETE, COPY, STACK, UNSTACK, CODE, SORT. For example:

LET c1(3) = 1.3	– changes 3rd element of c1 to 1.3
INSERT 9 10 c41	– allows you to insert data between rows 9 and 10 of c41; you get the DATA> prompt; finish with END
INSERT c41	– allows you to add data to end of c41
DELETE 2:9 c12	– deletes rows 2 to 9 of c12
COPY c1 c2	– copies c1 into c2 (there are several possible sub-commands to allow the inclusion or exclusion of specified rows)
STACK c1 c2 c3	– joins c2 onto c1 and puts into c3
SORT c1 c2	– puts elements of c1 in rank order in c2

Output

PRINT c2–c4 c7	– prints c2, c3, c4, c7 in columns
WRITE 'filename' c2–c5	– puts c2–c5 into a file called filename. This can be read in again using READ 'filename' c1–c4 for example. The filename must be kept exactly the same but the column numbers can be changed.

Save and retrieve

You can write the entire worksheet to a file with the SAVE command, e.g. SAVE 'DATA' – note quotation marks, but no column numbers. This file can be retrieved by RETR 'DATA'. When you list your files, you will find it called DATA.MTW. Do not try to print it as it is in binary and can only be read by RETR.

More on the LET command

If e denotes either a column or a constant, then the general form is:

LET e = arithmetic expression which may involve $+$, $-$, $*$, $/$, $**$, brackets, and functions such as sqrt, loge, sin, absolute value (or abso), etc.

For example:

$$\text{LET } c5 = (c1 - \text{MEAN}(c1))$$
$$\text{LET } k2 = 3*k1 + \text{MEAN}(c1)$$

More commands

k denotes a constant (e.g. k7), c a column, and e either. Square brackets denote an optional argument. Commas are included in some commands but are optional. There should be at least one space between each item.

ERASE c3–c5	– erases c3, c4, c5
RESTART	– erases the whole worksheet
INFORMATION	– gives current status of worksheet including all columns in use

Functions

FUNC e, e	– evaluates the function of e and puts into e. Available functions include SQRT (square root), LOGT (log to base 10), LOGE, ANTILOG, EXPO, ROUND, SIN, COS, TAN, ASIN, NSCORE, etc.: it is often easier to use the LET command, e.g. LET c1 = SQRT (c2)

Column operation

COUNT c1, [k1]   – counts number of elements in c1 and (optionally) puts the number into k1

Other commands of similar form are MEAN, SUM, STDEV (standard deviation), MEDI(an), MAXI(mum), MINI(mum). The DESCRIBE command is also very useful. It gives summary statistics, e.g. DESCRIBE c1.

Tables

Data must be integer-valued.

TALLY c1   – prints a discrete frequency distribution
TABLE c3 c4   – gives a two-way table
TABLE c,..., c   – gives a multi-way table

Graphs

HIST c1   – histogram of c1
STEM c5   – stem-and-leaf plot
BOXP c7   – boxplot
BOXP c7;   – box plot of c7 at each level of c9; c9 must be discrete and
BY c9.   same length as c7
DOTPLOT c1   – dotplot
PLOT c1 c3   – plots a scatter diagram of the observations in c1 against those in c3

There are plotting commands to adjust the shape of graphs from the default shape and to provide flexible ways of specifying scales. High-resolution graphics are also available.

Naming columns

It is often helpful to give columns names, e.g. name c1 'length' – thereafter you can refer to c1 as c1 or 'length', but you must include the quotation marks.

Distributions and random numbers

There are four commands dealing with various statistical distributions. RAND generates random numbers from a specified distribution into one or more columns. PDF computes pdf; CDF computes cdf and INVCDF computes inverse of a distribution function. The allowed distributions include Bernoulli, binomial, Poisson, discrete uniform, uniform, normal, t, F, χ^2, Cauchy, exponential, gamma, Weibull, beta, lognormal, logistic. For example:

RAND k c1;
BERNOULLI 0.6.

— generates k zero-one values with Prob(one) = 0.6

RAND k c7;
NORMAL 35 2.

— generates k $N(35, 2^2)$ observations and puts into c7

PDF;
BINO n = 10 p = 0.4.

— prints binomial probabilities for $n = 10$, $p = 0.4$
(Note: if n is large and $p > \frac{1}{2}$, use $q = 1 - p$ for $(n - X)$ or use Poisson approximation.)

PDF;
POIS 7.

— prints Poisson probabilities for Pois (mean = 7)
(Note: mean must be < 100 and preferably much smaller.)

RAND k c1;
UNIF a b.

— random numbers from uniform distribution on (a, b)

For continuous distributions (or discrete), pdf calculates pdf at values stored in cx, say, and puts them into cy, e.g. pdf cx cy; followed by definition of distribution.

Estimation and hypothesis testing

ZINT [k_1] k_2 c8

— k_1% CI for μ assuming sigma = k_2; data in c8; $k_1 = 95$ is default

TINT [k] c7

— k% CI for μ, sigma unknown

ZTES k_1 k_2 c8

— normal test of mean = k_1 assuming sigma = k_2

TTES k_1 c6

— t-test of mean = k_1

Both ZTES and TTES assume two-sided alternative. If $H_1:\mu < k_1$ use subcommand ALTE = $- 1$. If $H_1:\mu > k_1$, use subcommand ALTE = $+ 1$.

TWOS [k] c1, c2

— two-sample t-test and k% CI for difference in means. Data in c1 and c2. The subcommand ALTERNATIVE can be used for a one-sided test. The subcommand POOLED can be used if a pooled estimate of common variance is to be calculated.

Regression

REGR c5 2 c7 c11 [c20 c21]

— does a multiple regression of c5 on two explanatory variables, c7 and c11. All columns must have same length. Optionally put standardized residuals in c20 and fitted values in c21. There are many possible subcommands.

CORR c1 c2

— calculates correlation coefficient

Matrices

READ 3 4 m2	– reads (3 × 4) matrix into m2; data must be three rows of four observations
PRIN m2	– prints m2
INVE m2 m3	– inverts m2 and puts it into m3; m2 must be square and non-singular
TRAN m2 m3	– transposes m2; puts into m3
ADD m1 m2 m3	– add m1 to m2, put into m3; similarly for SUBT and MULT
EIGEN m1 c1 m2	– calculates eigenvalues and vectors of (symmetric) m1, put into c1 and m2

ANOVA

AOVO c1 c2 c3	– one-way ANOVA of three groups of observations; group 1 observations in c1, etc.; or equivalently use ONEW on stacked data e.g
ONEW c1 c2	– data in c1, corresponding elements of c2 denote group numbers
TWOW c1 c2 c3 [c4, c5]	– two-way ANOVA of c1 data; block numbers in c2, treatment numbers in c3; optionally put residuals into c4 and fitted values into c5

Time series

If data are a time series, then:

ACF c2	– calculates acf of data in c2
PACF c2	– partial acf
DIFF 1 c2 c4	– puts first differences of c2 into c4
ARIM 2 1 1 c2	– fits an ARIMA (2, 1, 1) model to the data in c2
ARIM p d q, P D Q, S, c2	– fits seasonal ARIMA model; S = season length
TSPLOT c2	– time-series plot of c2

Other options

CHIS c1 c2 c3	χ^2 test
MANN c1 c2	Mann–Whitney test
RANK c1 c3 c5	calculates ranks for elements in each column

There are many other commands and subcommands. You should (a) consult reference manual (b) use HELP commands, or (c) guess (!). Many commands are obvious, e.g. to stop outlying values being trimmed from a stem-and-left plot, try:

STEM c5;
NOTRIM.

Macros

Sequences of MINITAB commands, called 'exec macros', can be stored in a text file, and then executed using the EXECUTE command.

Logging

You can log a session with the PAPER, or OUTFILE 'filename', commands.

Example B.1.

The MINITAB analysis for the data in Exercise E.1 is as follows.

```
SET c1
35.2 57.4 27.2   .   .   .   (read data in three rows of 10 observations)
.   .   .   .   .   .   .   .
.   .   .   .   .   44.4
END
SET c2
(1,2,3)8                     (gives 1 repeated eight times, then 2 repeated eight
END                         times, 3 repeated eight times)
SET c3
3(1,2,3,4,5,6,7,8)          (gives 1 to 8 repeated three times)
END
TWOW c1,c2,c3,c4            (gives two-way ANOVA; put residuals into c4)
PRIN c4
etc.
```

B.2 Glim

GLIM is a powerful program for fitting generalized linear models. The user has to specify the response variable, the error distribution, the link function and the predictor variables. These notes describe Release 3.77, though Release 4 is now available (see Francis, Green and Payne, 1993).

Commands

All commands start with a $. A command may consist of several letters but only the first three are recognized. Commands are also terminated by a $ but you do not usually need to type this as the initial $ of the next command also serves to terminate the previous command. The exception is when you want a command to be implemented straight away. For example, $STOP$ exits you from the package. It will take too long to explain all commands but many are self-explanatory.

Vectors

The observations and explanatory variables are stored in vectors. User-defined vectors consist of a letter followed optionally by further letters and/or digits. Upper- and lower-case letters are interchangeable. Only the first four are recognized. A VARIATE can contain any real number while a FACTOR contains integers from the set $\{1, 2, \ldots, k\}$ for a specified k. A system-defined vector consists of the function symbol (%) followed by two letters.

Scalars

These store a single number. They are denoted by % followed by a single letter, such as %A. Only 26 are available and they do not need to be declared before use. System-defined scalars, such as %GM, consists of % followed by two letters.

Functions

The symbol % also denotes a function as in %EXP, %LOG, %SIN, %SQRT, etc.

Reading in data

The UNITS command sets a standard length for all vectors. Variates and factors may be declared by the VAR and FAC commands, e.g. $FAC 20 A 3$ declares a factor length 20 (not needed if $UNITS 20$ has been declared) with three levels. The DATA command specifies the set of identifiers for the vectors whose values you wish to assign and which will be read by the next READ statement. The READ command then reads the data. Use the LOOK command to check the data, and the CALC command to correct errors (e.g. $CALC X(4) = 19.7$ assigns new value 19.7 to the fourth element of the X-vector). Data can also be read from a file. The examples below explain these commands more easily.

Other commands

$PLOT X Y	plots a scatter diagram
$CALC X = Y + Z	forms a new vector
$CALC X = X**2 + 1	a transformation
$CALC X = %GL(k, n)	useful for assigning factor values. This gives the integers 1 to k in blocks of n, repeated until the UNITS length is achieved. For example if UNITS = 14, %GL(3,2) gives 1,1,2,2,3,3,1,1,2,2,3,3,1,1.

Defining the model

The YVARIATE command specifies the response variable. The ERROR command specifies the error distribution which may be normal (N), Poisson (P), binomial (B) or gamma (G). For the binomial case the name of the sample size variable, in the proportion r/n, must also be specified (e.g. $ERROR B N). The LINK command specifies the link function (e.g. I for identity, L for log, G for logit, R for reciprocal, P for probit, S for square root, E for exponent, together with the value of the exponent). Only meaningful combinations of ERROR and LINK are allowed, (e.g. binomial with probit or logit). The FIT command fits a model with specified predictor variables (e.g. $FIT Z W). $FIT A*B fits factors A and B with interaction. $FIT + A adds factor A to the previously fitted model.

Displaying the results

The DISPLAY command produces results of the previous fit. The scope of the output is specified by letters which include:

E – estimates of parameters
R – fitted values and residuals
V – variances of estimates
C – correlation matrix of parameter estimates.

The program also produces the following variables which you can display (using LOOK), use in calculations, or store for future reference:

%DF – degrees of freedom for error
%DV – the deviance
%FV – fitted values in vector of standard length
etc.

Example B.2

The GLIM analysis for the data in Exercise E.1 is as follows:

```
$UNITS 24 $FAC COL 8 ROW 3
$DATA OBS $READ
35.2 57.4.   .   .
  .   .   .   .   .   .              (three rows of eight observations)
  .   .   .   .   44.4
$CALC ROW = %GL(3,8):COL = %GL(8,1)
$YVAR OBS
$ERR N
$LINK I
$FIT $                              (fits null model; deviance is total
                                     corrected SS)
$FIT ROW: +COL$                     (fits row terms and then adds
                                     column effects)

$DISPLAY ER $
$CALC E = OBS − %FV$                (calculates residuals)
$LOOK E $                           (tabulates residuals)
etc.
```

The ANOVA table in its usual form can be obtained from the deviances produced by GLIM by appropriate subtractions. You may prefer the output from the MINITAB package in this case – see Example B.1 above.

Example B.3

The GLIM analysis for the data in Exercise E.2 is (in brief) as follows:

```
$UNITS 9 $FAC DAY 3 ENG 3 BURN 3
$DATA Y BURN $READ
16 1                                (observations in row order
17 2                                 with corresponding burner number)

 .  .
13 2
$CALC DAY = %GL (3,3)
$CALC ENG = %GL (3,1)
$YVAR Y $ERR N $LINK I
$FIT $
$FIT DAY $
$FIT + ENG $
$FIT + BURN $
$DISPLAY ER$
etc.
```

Example B.4

The GLIM analysis for the data in Exercise G.5 is (in brief) as follows:

```
$UNITS 7
$DATA X N R
$READ
0.9  46  17
1.1  72  22
 .   .   .                    (seven lines of data)
4.0  38  30
$YVAR R $ERR B N $LINK G$     (this specifies logit link;
                              also try P for probit)

$FIT $
$FIT X $
$DISPLAY ER$
$PLOT R %FV$                  (plots r against fitted values)
$CALC P = R/N $               (generates new proportion variable)
etc.
```

Appendix C
Some useful addresses

The practising statistician should consider joining one of the many national and international statistical societies. Potential benefits may include (a) access to a library, (b) the receipt of up-to-date journals, (c) a regular newsletter giving details of forthcoming lectures and conferences, and (d) the opportunity to meet other statisticians. Most countries have a national statistical society and the relevant address can probably be obtained via the statistics department of your local university or by writing to:

International Statistical Institute,
428 Prinses Beatrixlaan,
PO Box 950,
2270 AZ Voorburg,
Netherlands.

Full membership of this international society (the ISI) is by election only, but there are many sections and affiliated organizations which are open to all. The ISI publishes an annual directory of statistical societies. The sections include:

Bernoulli Society for Mathematical Statistics and Probability
The International Association for Statistical Computing
The International Association of Survey Statisticians

The affiliated organizations include many national societies such as:

Statistical Society of Australia
Statistical Society of Canada
Swedish Statistical Association

There are also some general, or special-interest, societies of a more international character which are also mostly affiliated to the ISI. They include the following (note that the addresses are correct at the time of writing, but it is advisable to check where possible):

American Statistical Association, 1429 Duke St., Alexandria, VA 22314-3402, USA

American Society for Quality Control, 611 E. Wisconsin Ave., PO Box 3005, Milwaukee, WI 53201-3005, USA

Royal Statistical Society, 12 Errol St., London EC1, UK (Note: since January 1993, the Royal Statistical Society has merged with the Institute of Statisticians)

International Biometric Society, Suite 401, 1429 Duke St., Alexandria, VA 22314-3402, USA

Institute of Mathematical Statistics, Business Office, 3401 Investment Boulevard #7, Hayward, CA 94545, USA

Indian Statistical Institute, 203 Barrackpore Trunk Road, Calcutta 700 035, India

Appendix D
Statistical tables

Most textbooks provide the more commonly used statistical tables including percentage points of the normal, t-, χ^2- and F-distributions. In addition, there are many more comprehensive sets of tables published including:

Fisher, R. A. and Yates, F. (1963) *Statistical Tables for Biological, Agricultural and Medical Research*, 6th edn, Oliver and Boyd, London.
Neave, H. R. (1978) *Statistics Tables*, George Allen and Unwin, London.
Pearson, E. S. and Hartley, H. O. (1966) *Biometrika Tables for Statisticians*, Vol. 1, 3rd edn, Cambridge University Press, Cambridge.

There are also many specialized sets of tables, including, for example, tables of the binomial probability distribution, and numerous tables relating to quality control.

A useful collection of mathematical functions and tables is:

Abramowitz, M. and Stegun, I. A. (1965) *Handbook of Mathematical Functions*, Dover, New York.

The following tables give only an abridged version of the common tables, as in my experience crude linear interpolation is perfectly adequate in most practical situations. The tables also give fewer decimal places than is often the case, because quoting an observed F-value, for example, to more than one decimal place usually implies spurious accuracy.

Table D.1.1 Areas under the standard
normal curve

z	Prob (obs. $> z$) $\times 100$
0.0	50
0.5	30.9
1.0	15.9
1.28	10.0
1.5	6.7
1.64	5.0
1.96	2.5
2.33	1.0
2.57	0.5
3.0	0.14
3.5	0.02

The tabulated values show the percentage of observations which exceed the given value, z, for a normal distribution, mean zero and standard deviation one; thus the values are one-tailed.

Table D.1.2 Percentage points of Student's t-distribution

	Two-tailed probabilities			
	0.10	0.05	0.02	0.01
	One-tailed probabilities (α)			
v	0.05	0.025	0.01	0.005
1	6.34	12.71	31.82	63.66
2	2.92	4.30	6.96	9.92
3	2.35	3.18	4.54	5.84
4	2.13	2.78	3.75	4.60
6	1.94	2.45	3.14	3.71
8	1.86	2.31	2.90	3.36
10	1.81	2.23	2.76	3.17
15	1.75	2.13	2.60	2.95
20	1.72	2.09	2.53	2.84
30	1.70	2.04	2.46	2.75
60	1.67	2.00	2.39	2.66
∞	1.64	1.96	2.33	2.58

The one-tailed values $t_{\alpha,v}$, are such that $\text{Prob}(t_v > t_{\alpha,v}) = \alpha$ for Student's t-distribution on v degrees of freedom. The two-tailed values are such that $\text{Prob}(|t_v| > t_{\alpha,v}) = 2\alpha$, since the t-distribution is symmetric about zero. Interpolate for any value of v not shown above.

Table D.1.3 Percentage points of
the χ^2-distribution

ν	α	
	0.05	0.01
1	3.8	6.6
2	6.0	9.2
3	7.8	11.3
4	9.5	13.3
6	12.6	16.8
8	15.5	20.1
10	18.3	23.2
12	21.0	26.2
14	23.7	29.1
16	26.3	32.0
20	31.4	37.6
25	37.6	44.3
30	43.8	50.9
40	55.8	63.7
60	79.1	88.4
80	101.9	112.3

The values $\chi^2_{\alpha,\nu}$ are such that $\mathrm{Prob}(\chi^2_\nu > \chi^2_{\alpha,\nu}) = \alpha$ for the χ^2-distribution on ν degrees of freedom. The χ^2-distribution is not symmetric, but the lower percentage points are rarely needed and will not be given here. Note that $E(\chi^2_\nu) = \nu$. Interpolate for any value of ν not shown above or use an appropriate approximation. For large ν, the χ^2-distribution tends to $N(\nu, 2\nu)$, but a better approximation can be obtained using the fact that $[\sqrt{(2\chi^2_\nu)} - \sqrt{(2\nu - 1)}]$ is approximately $N(0, 1)$.

Table D.1.4 Percentage points of the F-distribution

(a) 5% values ($\alpha = 0.05$)

v_2 \ v_1	1	2	4	6	8	10	15	30	∞
1	161	199	225	234	239	242	246	250	254
2	18.5	19.0	19.2	19.3	19.4	19.4	19.4	19.5	19.5
3	10.1	9.5	9.1	8.9	8.8	8.8	8.7	8.6	8.5
4	7.7	6.9	6.4	6.2	6.0	6.0	5.9	5.7	5.6
5	6.6	5.8	5.2	4.9	4.8	4.7	4.6	4.5	4.4
6	6.0	5.1	4.5	4.3	4.1	4.1	3.9	3.8	3.7
8	5.3	4.5	3.8	3.6	3.4	3.3	3.2	3.1	2.9
10	5.0	4.1	3.5	3.2	3.1	3.0	2.8	2.7	2.5
12	4.7	3.9	3.3	3.0	2.8	2.7	2.6	2.5	2.3
15	4.5	3.7	3.1	2.8	2.6	2.5	2.4	2.2	2.1
20	4.3	3.5	2.9	2.6	2.4	2.3	2.2	2.0	1.8
30	4.2	3.3	2.7	2.4	2.3	2.2	2.0	1.8	1.6
40	4.1	3.2	2.6	2.3	2.2	2.1	1.9	1.7	1.5
∞	3.8	3.0	2.4	2.1	1.9	1.8	1.7	1.5	1.0

(b) 1% values ($\alpha = 0.01$)

v_2 \ v_1	1	2	4	6	8	10	15	30	∞
1	4050	5000	5620	5860	5980	6060	6160	6260	6370
2	98.5	99.0	99.2	99.3	99.4	99.4	99.4	99.4	99.5
3	34.1	30.8	28.7	27.9	27.5	27.2	26.9	26.5	26.1
4	21.2	18.0	16.0	15.2	14.8	14.5	14.2	13.8	13.5
5	16.3	13.3	11.4	10.7	10.3	10.0	9.7	9.4	9.0
6	13.7	10.9	9.1	8.5	8.1	7.9	7.6	7.2	6.9
8	11.3	8.6	7.0	6.4	6.0	5.8	5.5	5.2	4.9
10	10.0	7.6	6.0	5.4	5.1	4.8	4.6	4.2	3.9
12	9.3	6.9	5.4	4.8	4.5	4.3	4.0	3.7	3.4
15	8.7	6.4	4.9	4.3	4.0	3.8	3.5	3.2	2.9
20	8.1	5.8	4.4	3.9	3.6	3.4	3.1	2.8	2.4
30	7.6	5.4	4.0	3.5	3.2	3.0	2.7	2.4	2.0
40	7.3	5.2	3.8	3.3	3.0	2.8	2.5	2.2	1.8
∞	6.6	4.6	3.3	2.8	2.5	2.3	2.0	1.7	1.0

The values F_{α, v_1, v_2} are such that $\text{Prob}(F_{v_1, v_2} > F_{\alpha, v_1, v_2}) = \alpha$ for an F-distribution with v_1 (numerator) and v_2 (denominator) degrees of freedom. The F-distribution is not symmetric and lower percentage points can be found using $F_{1-\alpha, v_1, v_2} = 1/F_{\alpha, v_2, v_1}$, where the order of the degrees of freedom is reversed. Interpolate for any values of v_1, v_2 not shown above.

References

Aitkin, M. and Clayton, D. (1980) The fitting of exponential, Weibull and extreme value distributions to complex censored survival data using GLIM. *Appl. Stat.*, **29**, 156–63.

Altman, D. G. (1991) *Practical Statistics for Medical Research*, Chapman and Hall, London.

Altman, D. G., Gore, S. M., Gardner, M. J. and Pocock, S. J. (1983) Statistical guidelines for contributors to medical journals. *Br. Med. J.* **286**, 1489–93.

Andersen, B. (1990) *Methodological Errors in Medical Research*, Blackwell Scientific, Oxford.

Anderson, C. W. and Loynes, R. M. (1987) *The Teaching of Practical Statistics*, Wiley, Chichester.

Anscombe, F. J. (1973) Graphs in statistical analysis. *Am. Statn.*, **27**, 17–21.

Armstrong, J. S. (1985) *Long-Range Forecasting*, 2nd edn, Wiley, New York.

Atkinson, A. C. (1985) *Plots, Transformations and Regression*, Oxford University Press, Oxford.

Bain, L. J. and Engelhardt, M. (1987) *Introduction to Probability and Mathematical Statistics*, 2nd edn, PWS-Kent, Boston.

Barnard, G. A. (1986) Rescuing our manufacturing industry – some of the statistical problems. *The Statistician*, **35**, 3–16.

Barnett, V. (1982) *Comparative Statistical Inference*, 2nd edn, Wiley, Chichester.

Barnett, V. (1991) *Sample Survey Principles and Methods*, 2nd edn, Edward Arnold, London.

Barnett, V. and Lewis, T. (1985) *Outliers in Statistical Data*, 2nd edn, Wiley, Chichester.

Bates, D. M. and Watts, D. G. (1988) *Non-linear Regression and its Applications*, Wiley, New York.

Becker, R. A., Cleveland, W. S. and Wilks, A. R. (1987) Dynamic graphics for data analysis. *Stat. Science*, **2**, 355–395.

Begg, C. B. and Berlin, J. A. (1988) Publication bias: a problem in interpreting medical data (with discussion). *J. R. Stat. Soc. A*, **151**, 419–463.

Bernardo, J. M. and Smith, A. F. M. (1994) *Bayesian Theory*, Wiley, Chichester.

Bissell, A. F. (1989) Interpreting mean squares in saturated fractional designs. *J. App. Statistics*, **16**, 7–18.

Bissell, A. F. (1994) *Statistical Methods for SPC and TQM*, Chapman and Hall, London.

de Bono, E. (1967) *The Use of Lateral Thinking*, Jonathan Cape, London. (Republished by Penguin Books.)

Box, G. E. P. (1976) Science and statistics. *J. Amer. Stat. Assoc.*, **71**, 791–799.

Box, G. E. P. (1980) Sampling and Bayes' inference in scientific modelling and robustness (with discussion). *J. R. Stat. Soc. A*, **143**, 383–430.

Box, G. E. P. (1983) An apology for ecumenism in statistics, in *Scientific Inference, Data Analysis and Robustness* (eds G. E. P. Box, T. Leonard and C. F. Wu), Academic Press, New York.

Box, G. E. P. (1990) Commentary on a paper by Hoadley and Kettenring. *Technometrics*, **32**, 251–252.

Box, G. E. P. (1994) Statistics and quality improvement. *J. R. Stat. Soc. A*, **157**, 209–229.

Box, G. E. P., Hunter, W. G. and Hunter, J. S. (1978) *Statistics for Experimenters*, Wiley, New York.

Box, G. E. P., Jenkins, G. M. and Reinsel, G. C. (1994) *Time Series Analysis, Forecasting and Control*, 3rd edn, Prentice Hall, Englewood Cliffs, NJ.

Carver, R. (1978) The case against statistical significance testing. *Harv. Ed. Rev.* **48**, 378–99.

Chambers, J. M. (1993) Greater or lesser statistics: a choice for future research. *Statistics and Computing*, **3**, 182–184.

Chambers, J. M., Cleveland, W. S., Kleiner, B. and Tukey, P. A. (1983) *Graphical Methods for Data Analysis*, Wadsworth, Belmont, Calif.

Chapman, M. (1986) *Plain Figures*, HMSO, London.

Chatfield, C. (1982) Teaching a course in applied statistics. *Appl. Stat.* **31**, 272–89.

Chatfield, C. (1983) *Statistics for Technology*, 3rd edn, Chapman and Hall, London.

Chatfield, C. (1985) The initial examination of data (with discussion). *J. R. Stat. Soc. A*, **148**, 214–53.

Chatfield, C. (1986) Exploratory data analysis. *Eur. J. Op. Res.*, **23**, 5–13.

Chatfield, C. (1989) *The Analysis of Time Series*, 5th edn, Chapman and Hall, London.

Chatfield, C. (1991) Avoiding statistical pitfalls (with discussion). *Stat. Science*, **6**, 240–268.

Chatfield, C. (1995) Model uncertainty, data mining and statistical inference. *J. R. Stat. Soc.* (to appear).

Chatfield, C. and Collins, A. J. (1980) *Introduction to Multivariate Analysis*, Chapman and Hall, London.

Cleveland, W. S. (1985) *The Elements of Graphing Data*, Wadsworth, Belmont, Calif.

Cleveland, W. S. (1993) *Visualizing Data*, Hobart Press, Summit, NJ.

Cleveland, W. S. and McGill, R. (1987) Graphical perception: the visual decoding of quantitative information on graphical displays of data. *J. R. Stat. Soc. A*, **150**, 192–229.

Cleveland, W. S., Diaconis, P. and McGill, R. (1982) Variables on scatterplots look more highly correlated when the scales are increased. *Science*, **216**, 1138–41.

Cochran, W. G. (1977) *Sampling Techniques*, 3rd edn, Wiley, New York.

Cochran, W. G. (1983) *Planning and Analysis of Observational Studies*, Wiley, New York.

Cochran, W. G. and Cox, G. M. (1957) *Experimental Designs*, 2nd edn, Wiley, New York.

Collett, D. (1994) *Modelling Survival Data in Medical Research*, Chapman and Hall, London.

Cook, D. and Weisberg, S. (1982) *Residuals and Influence in Regression*, Chapman and Hall, London.

Cox, D. R. (1958) *The Planning of Experiments*, Wiley, New York.

Cox, D. R. (1977) The role of significance tests. *Scand. J. Stat.*, **4**, 49–70.

Cox, D. R. (1981) Theory and general principles in statistics. *J. R. Stat. Soc. A*, **144**, 289–97.

Cox, D. R. (1986) Some general aspects of the theory of statistics. *Int. Stat. Rev.* **54**, 117–26.

Cox, D. R. (1990) Role of statistical models in statistical analysis. *Stat. Science*, **5**, 169–174.

Cox, D. R. and Oakes, D. (1984) *Analysis of Survival Data*, Chapman and Hall, London.

Cox, D. R. and Snell, E. J. (1981) *Applied Statistics*, Chapman and Hall, London.

Cramér, H. (1946) *Mathematical Methods of Statistics*, Princeton University Press, Princeton, NJ.

Crowder, M. J., Kimber, A. C., Smith, R. L. and Sweeting, T. J. (1991) *Statistical Analysis of Reliability Data*, Chapman and Hall, London.

Daniel, C. and Wood, F. S. (1980) *Fitting Equations to Data*, 2nd edn, Wiley, New York.

Dekker, A. L. (1994) Computer methods in population census data processing. *Int. Stat. Review*, **62**, 55–70.

Deming, W. E. (1986) *Out of the Crisis: Quality, Productivity and Competitive Position*, MIT, Center for Advanced Engineering Study, Cambridge, Mass.; and Cambridge University Press, Cambridge.

Diaconis, P. and Efron, B. (1985) Testing for independence in a two-way table: new interpretations of the chi-square statistic (with discussion). *Ann. Stat.*, **13**, 845–74.

Diamond, M. and Stone, M. (1981) Nightingale on Quetelet: Part I. *J. R. Stat. Soc. A*, **144**, 66–79.

Dineen, J. K., Gregg, P. and Lascelles, A. K. (1978) The response of lambs to vaccination at weaning with irradiated *Trichostrongylus colubriformis* larvae: segregation into 'responders' and 'non-responders'. *Int. J. Parasit.* **8**, 59–63.

Dobson, A. J. (1990) *An Introduction to Generalized Linear Models*, 2nd edn, Chapman and Hall, London.

Draper, D. (1995) Assessment and propagation of model uncertainty (with discussion). *J. R. Stat. Soc. B*, **57**, 45–97.

Draper, N. R. and Smith, H. (1981) *Applied Regression Analysis*, 2nd edn, Wiley, New York.

Draper, D. *et al.* (1992) *Combining Information*, National Academy Press, Washington, DC (republished 1993 by Amer. Statist. Assoc. as Number 1 in Contemporary Statistics series).

Duncan, A. J. (1974) *Quality Control and Industrial Statistics*, 4th edn, Irwin, Homewood, Ill.

Edwards, D. and Hamson, M. (1989) *Guide to Mathematical Modelling*, Macmillan Educational Ltd, London.

Efron, B. and Gong, G. (1983) A leisurely look at the bootstrap, the jackknife and cross-validation. *Am. Statn*, **37**, 36–48.

Efron, B. and Tibshirani, R. J. (1993) *An Introduction to the Bootstrap*, Chapman and Hall, New York.

Ehrenberg, A. S. C. (1982) *A Primer in Data Reduction*, Wiley, Chichester.

Ehrenberg, A. S. C. (1984) Data analysis with prior knowledge, in *Statistics: An Appraisal* (eds H. A. David and H. T. David), Iowa State University Press, Ames, Iowa, pp. 155–82.

Ehrenberg, A. S. C. and Bound, J. A. (1993) Predictability and prediction (with discussion). *J. R. Stat. Soc. A*, **156**, 167–206.

Erickson, B. H. and Nosanchuk, T. A. (1992) *Understanding Data*, 2nd edn, Open University Press, Milton Keynes.

Everitt, B. S. and Dunn, G. (1991) *Advanced Multivariate Data Analysis*, 2nd edn, Arnold, London.

Feller, W. (1968) *An Introduction to Probability Theory and its Applications*, 3rd edn, Wiley, New York.

Feynman, R. P. (1985) *Surely You're Joking Mr Feynman*, Unwin Paperbacks, London.

Foreman, E. (1991) *Survey Sampling Principles*, Marcel Dekker, Berlin.

Francis, B., Green, M. and Payne, C. (eds.) (1993) *The GLIM System: Release 4 Manual*, Clarendon Press, Oxford.

Friedman, L. M., Furberg, C. D. and Demets, D. L. (1985) *Fundamentals of Clinical Trials*, 2nd edn, PSG Publishing, Littleton, Mass.

Geisser, S. (1993) *Predictive Inference: an Introduction*, Chapman and Hall, New York.

Gilchrist, W. (1984) *Statistical Modelling*, Wiley, Chichester.

Gnanadesikan, R. (1977) *Methods for Statistical Data Analysis of Multivariate Observations*, Wiley, New York.

Goodhardt, G. J., Ehrenberg, A. S. C. and Chatfield, C. (1984) The Dirichlet: a comprehensive model of buying behaviour (with discussion). *J. R. Stat. Soc. A*, **147**, 621–55.

Gore, S. M. and Altman, D. G. (1982) *Statistics in Practice*, British Medical Association, London.

Gore, S. M., Jones, I. G. and Rytter, E. C. (1977) Misuse of statistical methods: critical assessment of articles in BMJ from January to March 1976. *Brit. Medical J.*, **1**, 85–87.

Gower, J. C. (1993) The next ten years in statistics? *Statistics and Computing*, **3**, 191–193.

Gowers, Sir Ernest (1986) *The Complete Plain Words*, 3rd edn, Penguin, Harmondsworth.

Granger, C. W. J. and Newbold, P. (1986) *Forecasting Economic Time Series*, 2nd edn, Academic Press, New York.

Green, P. J. and Chatfield, C. (1977) The allocation of university grants. *J. R. Stat. Soc. A*, **140**, 202–9.

Green, P. J. and Silverman, B. W. (1993) *Nonparametric Regression and Generalized Linear Models*, Chapman and Hall, London.

Greenacre, M. (1984) *Theory and Applications of Correspondence Analysis*, Academic Press, London.

Hahn, G. J. (1984) Experimental design in the complex world. *Technometrics*, **26**, 19–31.

Hamaker, H. C. (1983) Teaching applied statistics for and/or in industry, in *Proceedings of the First International Conference on Teaching Statistics* (eds D. R. Grey, P. Holmes, V. Barnett and G. M. Constable), Teaching Statistics Trust, Sheffield, pp. 655–700.

Hand, D. J. (ed.) (1993) *Artificial Intelligence Frontiers in Statistics*, Chapman and Hall, London.

Hand, D. J. (1994) Statistical expert systems. *Chance*, **7**, 28–34.

Hand, D. J. and Everitt, B. S. (eds) (1987) *The Statistical Consultant in Action*, Cambridge University Press, Cambridge.

Hand, D. J., Daly, F., Lunn, A. D., McConway, K. J. and Ostrowski, E. (eds.) (1994) *A Handbook of Small Data Sets*, Chapman and Hall, London.

Hawkes, A. G. (1980) Teaching and examining applied statistics. *The Statistician*, **29**, 81–9.

Healey, D. (1980) *Healey's Eye*, Jonathan Cape, London.

Henderson, H. V. and Velleman, P. F. (1981) Building multiple regression models interactively. *Biometrics*, **37**, 391–411.

Hoaglin, D. C., Mosteller, F. and Tukey, J. W. (eds) (1983) *Understanding Robust and Exploratory Data Analysis*, Wiley, New York.

Hollander, M. and Proschan, F. (1984) *The Statistical Exorcist: Dispelling Statistics Anxiety*, Marcel Dekker, New York.

Hooke, R. (1983) *How to Tell the Liars from the Statisticians*, Marcel Dekker, New York.

Huff, D. (1959) *How to Take a Chance*, Pelican Books, London.

Huff, D. (1973) *How to Lie with Statistics*, 2nd edn, Penguin Books, London.

Hurlbert, S. H. (1984) Pseudoreplication and the design of ecological field experiments. *Ecological Monographs*, **54**, 187–211.

Inman, H. F. (1994) Karl Pearson and R. A. Fisher on statistical tests: a 1935 exchange from Nature. *Amer. Statistician*, **48**, 2–11.

International Statistical Institute (1986) Declaration on professional ethics. *Int. Stat. Rev.* **54**, 227–42.

Ishikawa, K. (1985) *What is Total Quality Control? The Japanese Way* (translated by D. J. Lu), Prentice Hall, Englewood Cliffs, NJ.

Janson, M. A. (1988) Combining robust and traditional least squares methods: a critical evaluation (with discussion). *J. Bus. & Econ. Stats.*, **6**, 415–451.

Johnson, N. L. and Kotz, S. (1969; 1971; 1972) *Distributions in Statistics* (four volumes), Wiley, New York (2nd edn of Vol. 1, *Discrete Distributions* with A. W. Kemp in 1992).

Joiner, B. L. (1982) Consulting, statistical, in *Encyclopedia of Statistical Sciences*, Vol. 2 (eds S. Kotz and N. L. Johnson), Wiley, New York, pp. 147–55.

Kanji, G. K. (1979) The role of projects in statistical education. *The Statistician*, **28**, 19–27.

Kendall, Sir Maurice and Ord, J. K. (1990) *Time Series*, 3rd edn, Edward Arnold, Sevenoaks, U.K.

Kimball, A. W. (1957) Errors of the third kind in statistical consulting. *J. Amer. Stat. Assoc.*, **52**, 133–142.

Kish, L. (1965) *Survey Sampling*, Wiley, New York.

Krzanowski, W. J. (1988) *Principles of Multivariate Analysis*, Clarendon Press, Oxford.

Lehmann, E. L. (1993) The Fisher, Neyman–Pearson theories of testing hypotheses: one theory or two? *J. Amer. Stat. Assoc.*, **88**, 1242–1249.

Lessler, J. T. and Kalsbeek, W. D. (1992) *Nonsampling Error in Surveys*, Wiley, New York.

Lindsay, R. M. and Ehrenberg, A. S. C. (1993) The design of replicated studies. *Am. Statn.*, **47**, 217–228.

Little, R. J. A. and Rubin, D. B. (1987) *Statistical Analysis with Missing Data*, Wiley, New York.

Lovell, M. C. (1983) Data mining, *Rev. Econ. and Stat.*, **65**, 1–12.

McCullagh, P. and Nelder, J. A. (1989) *Generalized Linear Models*, 2nd edn, Chapman and Hall, London.

McNeil, D. R. (1977) *Interactive Data Analysis*, Wiley, New York.

Manly, B. F. J. (1986) *Multivariate Statistical Methods*, Chapman and Hall, London.

Mantle, M. J., Greenwood, R. M. and Currey, H. L. F. (1977) Backache in pregnancy. *Rheumatology and Rehabilitation*, **16**, 95–101.

Mardia, K. V., Kent, J. T. and Bibby, J. M. (1979) *Multivariate Analysis*, Academic Press, London.

Meier, P. (1986) Damned liars and expert witnesses. *J. Am. Stat. Assoc.*, **81**, 269–76.

Miller, A. J. (1990) *Subset Selection in Regression*, Chapman and Hall, London.

Miller, R. G. Jr (1986) *Beyond ANOVA, Basics of Applied Statistics*, Wiley, Chichester.

Montgomery, D. C. (1991) *Design and Analysis of Experiments*, 3rd edn, Wiley, New York.

Morgan, B. J. T. (1984) *Elements of Simulation*, Chapman and Hall, London.

Morrison, D. E. and Henkel, R. E. (1970) *The Significance Test Controversy*, Butterworths, London.

Moser, C. A. (1980) Statistics and public policy. *J. R. Stat. Soc. A*, **143**, 1–31.

Moser, C. A. and Kalton, G. (1971) *Survey Methods in Social Investigation*, Heinemann, London.

Mosteller, F. (1988) Broadening the scope of statistics and statistics education, *Amer. Stat.*, **42**, 93–99.

Nelder, J. A. (1986) Statistics, science and technology: The Address of the President (with Proceedings). *J. R. Stat. Soc. A*, **149**, 109–21.

Oakes, M. (1986) *Statistical Inference: A Commentary for the Social and Behavioural Sciences*, Wiley, Chichester.

Patterson, H. D. and Silvey, V. (1980) Statutory and recommended list trials of crop varieties in the UK (with discussion). *J. R. Stat. Soc. A*, **143**, 219–52.

Penrose, R. (1989) *The Emperor's New Mind*, Oxford University Press, Oxford.

Peters, W. S. (1987) *Counting for Something: Principles and Personalities*, Springer-Verlag, New York.

Plewis, I. (1985) *Analysing Change: Measurement and Explanation using Longitudinal Data*, Wiley, Chichester.

Pocock, S. J. (1983) *Clinical Trials: A Practical Approach*, Wiley, Chichester.

Pólya, G. (1945) *How to Solve It*, Princeton University Press, Princeton, NJ.

Preece, D. A. (1981) Distributions of final digits in data. *The Statistician*, **30**, 31–60.

Preece, D. A. (1984) Contribution to the discussion of a paper by A. J. Miller. *J. R. Stat. Soc. A*, **147**, 419.

Preece, D. A. (1986) Illustrative examples: illustrative of what. *The Statistician*, **35**, 33–44.

Preece, D. A. (1987) Good statistical practice. *The Statistician*, **36**, 397–408.

Priestley, M. B. (1981) *Spectral Analysis and Time Series*, Vols 1 and 2, Academic Press, London.

Ratkowsky, D. A. (1983) *Non-Linear Regression Modelling*, Marcel Dekker, New York.

Reese, R. A. (1994) The global university. *RSS News*, **21**, No. 8, 9.

Ross, G. J. S. (1990) *Nonlinear Estimation*, Springer-Verlag, New York.

Rustagi, J. S. and Wolfe, D. A. (eds) (1982) *Teaching of Statistics and Statistical Consulting*, Academic Press, New York.

Ryan, B. F. and Joiner, B. L. (1994) *Mintab Handbook*, 3rd edn, Duxbury Press, Belmont, Cal.

Schnaars, S. P. (1989) *Megamistakes: Forecasting and the Myth of Rapid Technological Change*, Free Press, New York.

Schumacher, E. F. (1974) *Small is Beautiful*, Sphere Books, London.

Scott, J. F. (1976) Practical projects in the teaching of statistics at universities. *The Statistician*, **25**, 95–108.

Silverman, B. W. (1986) *Density Estimation*, Chapman and Hall, London.

Snedecor, G. W. and Cochran, W. G. (1980) *Statistical Methods*, 7th edn, Iowa State University Press, Ames, Iowa.

Snee, R. D. (1974) Graphical display of two-way contingency tables. *Am. Statn*, **28**, 9–12.

Snell, E. J. (1987) *Applied Statistics: A Handbook of BMDP Analyses*, Chapman and Hall, London.

Snell, E. J. and Simpson, H. R. (1991) *Applied Statistics: A Handbook of GENSTAT Analyses*, Chapman and Hall, London.

Sprent, P. (1970) Some problems of statistical consultancy. *J. R. Stat. Soc. A*, **133**, 139–64.

Sprent, P. (1993) *Applied Nonparametric Statistical Methods*, 2nd edn, Chapman and Hall, London.

Steinberg, D. M. and Hunter, W. G. (1984) Experimental design: review and comment (with discussion). *Technometrics*, **26**, 71–130.

Stigler, S. M. (1986) *The History of Statistics*, Belknap Press of Harvard University Press, Cambridge, Mass.

Sundberg, R. (1994) Interpretation of unreplicated two-level factorial experiments. *Chemometrics and Intelligent Laboratory Systems*, **24**, 1–17.

Tam, S. M. and Green, B. W. (1994) A new methodology for processing Australia's 1991 population and housing census. *Int. Stat. Rev.*, **62**, 71–85.

Tufte, E. R. (1983) *The Visual Display of Quantitative Information*, Graphics Press, Cheshire, Conn.

Tukey, J. W. (1977) *Exploratory Data Analysis*, Addison-Wesley, Reading, Mass.

Unwin, A. (1992) How interactive graphics will revolutionize statistical practice. *The Statistician*, **41**, 365–369.

Velleman, P. F. and Hoaglin, D. C. (1981) *Applications, Basics, and Computing of Exploratory Data Analysis*, Duxbury Press, Boston, Mass.

Wainwright, G. (1984) *Report-Writing*. Management Update Ltd, London.

Wei, W. W. S. (1991) *Time Series Analysis*, Addison-Wesley, Redwood City, Calif.

Weisberg, S. (1985) *Applied Linear Regression*, 2nd edn, Wiley, New York.
Wetherill, G. B. (1982) *Elementary Statistical Methods*, 3rd edn, Chapman and Hall, London.
Wetherill, G. B. (1986) *Regression Analysis with Applications*, Chapman and Hall, London.

Author index

This index does not include entries in the reference section (pp. 311–317) where source references may be found. Where the text refers to an article or book by more than one author, only the first-named author is listed here.

Subject index